Plant Pests and their Control

Plant Pests

and their Control

Revised edition

Peter G. Fenemore, BSc, Dip Agr Sc, PhD

Reader in Entomology,
Department of Horticulture and Plant Health,
Massey University, Palmerston North, New Zealand

Butterworths
London Boston Durban Singapore Sydney Toronto Wellington

First published in New Zealand 1982
Revised UK edition, 1984

British Library Cataloguing in Publication Data

Fenemore, Peter G.
Plant pests and their control—Rev. British ed.
1. Pest control
I. Title
632'.9 SB950

ISBN 0-407-00304-5

Library of Congress Cataloguing in Publication Data

Fenemore, Peter G.
Plant pests and their control
Includes bibliographies and index.
1. Insect control. 2. Insects, Injurious and beneficial. 3. Plants, Protection of. I. Title.
SB931. F36 1983 632'.7 83-7504

ISBN 0-407-00304-5

Typeset by P. R. G. Graphics Ltd, Redhill
Printed and Bound in Great Britain by Whitstable Litho Ltd.

Foreword to the Revised Edition

In the past thirty years, major developments have occurred in the science of pest management, and the concepts of integrated pest control initiated by entomologists have been firmly accepted by pest, disease and weed control specialists throughout the world. Many specialised textbooks have emphasised the problems of over-reliance on conventional chemical pesticides, and highlighted the general principles and opportunities for integrating naturally occurring controls with more artificial methods. Much has been written on the ecological basis for pest management, and environment issues have been heavily stressed in the context of pesticide misuse.

Despite such widespread interest, integrated control, particularly in its more sophisticated forms, has relatively limited application. This is partly because it has tended to remain an ecological ideal and has not taken sufficiently into account the practical realities of pest, disease and weed control. In particular, inadequate attention has been given to pest management as part of the overall production system. Hence, practitioners, in part through ignorance, and in part through constraints of the production system, have been slow to adopt new pest management methods except where over-reliance on pesticides has caused major difficulties.

There is, therefore, urgent need to provide adequate understanding of the basis and practice of pest management for agriculturalists and for others concerned with overall problems of food and fibre production. In this respect Dr Peter Fenemore's book seems to me to provide an admirable basis for understanding the concepts of pest management and their application to control of insect and related pests within the framework of the production system, its constraints and opportunities. Moreover, the book provides a valuable introduction for those aiming to become specialists in particular aspects of pest management and will undoubtedly help to avoid a too narrow-minded viewpoint on the application of their disciplines.

M. J. Way
Professor of Applied Zoology
Imperial College
University of London

May 1983

Preface

Everyone engaged in agriculture or horticulture needs to know something about the pests that afflict the plants they cultivate, whether these be components of pasture, field crops, or horticultural plants for produce or ornamentation. In most cases this knowledge need not be exhaustive but should be sufficient to enable pest problems to be recognised and dealt with sensibly.

Pest control for the past several decades has relied heavily on the application of chemical insecticides. Valuable though these substances are, many problems have arisen in their use such as the appearance of resistant strains of pests, and harmful effects on non-target organisms. Most entomologists now agree that priority must be given to the development of pest management systems which harmoniously integrate various control measures (including insecticides), if pest control is to remain viable. Successful implementation of such pest management programmes requires considerably greater biological knowledge and skill on the part of practising agriculturalists and horticulturalists compared to the chemical era.

With this changing philosophy comes the recognition that insects are not by any means entirely harmful. They play important roles as pollinators of many cultivated plants and as natural enemies of pest species. In fact, the latter usually provide the cornerstone of effective pest management systems. An important aspect therefore of entomology as an applied science is not only control of harmful species but also the conservation and management of beneficial insects.

There is no doubt in the author's mind that successful manipulation of living organisms, whether harmful or beneficial, insects or otherwise, depends on adequate biological understanding. The object of this book is therefore to provide students, practising agriculturalists and horticulturalists, and other interested persons with a basic introduction to insects as living organisms and to the principles and practice of pest control. The treatment is introductory, not exhaustive, but should prove sufficient for understanding the basic concepts of modern pest control, including those of pest management. The foundation is also laid for further study in particular aspects for those that desire it.

P.G.F.

Massey University
Palmerston North

Acknowledgements

The author of any technical publication must rely heavily on information and assistance provided by others in the same and related fields, and this book is no exception.

In particular, I would like to acknowledge the help and encouragement provided by the following:

The Agricultural Chemicals Board,
Entomology Division, Department of Scientific and Industrial Research,
Plant Health and Diagnostic Station, Ministry of Agriculture and Fisheries,
Dr W. A. G. Charleston,
Professor K. S. Milne,
Dr G. W. Ramsay,
Mr E. Roberts,
Mr E.W. Valentine,
Dr C. H. Wearing,
Dr H. T. Wenham.

Irrespective of such assistance the final responsibility for the text and for any errors and omissions must remain the author's.

P.G.F.

ACKNOWLEDGEMENTS (Revised edition)

Agricultural Development and Advisory Service,
Pest Infestation Control Laboratory, Ministry of Agriculture, Fisheries and Food.
Central Veterinary Laboratory, Ministry of Agriculture, Fisheries and Food.

Contents

Chapter 1

Introduction

Since time immemorial mankind has had to contend with insect pests which attack his cultivated plants, his domestic animals and his person. Despite our modern technology this is no less true today than in times past. However, insects have also long provided man with valued products, in particular silk and honey, so their actions are not entirely harmful. In addition, in more recent years, the beneficial roles of insects in facilitating pollination of many cultivated plants and as natural enemies of pest species have been recognised. Some knowledge of insects, both harmful and beneficial, is therefore essential for anyone engaged in agriculture or horticulture.

What is entomology?

Entomology may be defined simply as the study of insects. This does not imply any aspect of economic significance and could involve no more than scientific curiosity. Such is not the intent of this book however which is concerned with the practical importance of insects, that is with *applied* or *economic* entomology as it may be properly termed.

Applied entomology has as its aim the manipulation of insects (both harmful and beneficial) to man's advantage. Successful manipulation of any organism depends on adequate biological understanding and this applies as much to insects as it does to cultivated plants or farm animals. Much of this book is therefore concerned with providing a general introduction to the biology and ecology of insects as a necessary prerequisite to their management.

What are insects?

Insects belong to that great subdivision of the Animal Kingdom called **Arthropoda** which are characterised by the possession of an external jointed skeleton encasing the body in a virtual suit of armour. Crayfish and crabs are familiar larger representatives of this group. The relationship between arthropods and other major subdivisions of the Animal Kingdom and the main subgroups of arthropods, including insects, are shown in Fig 1.

Like many other groups of animals, arthropods are segmented creatures; that is, the body is composed of a number of similar repeating units. In insects these body segments are grouped into three more or less distinct regions comprising head, thorax, and abdomen, as opposed to other classes of arthropods which have a different body plan. The main features which distinguish the major classes of arthropods are listed in Table 1.

Although entomology in the strict sense is concerned solely with insects the practising entomologist is expected to deal with related animals such as mites and spiders (or even the more remote slugs and snails), where these are of economic significance. There is thus some reference to such groups in this book.

The biological success of insects

If diversity of species, habits, numbers of individuals and persistence in geological time are any measure, then insects are undoubtedly one of the "success stories" of biological evolution.

Table 1. The main features which distinguish the major classes of arthropods

Class	CRUSTACEA (crayfish, crabs, woodlice, etc.) Head and thorax merged. Appendages on nearly all segments and modified as walking legs, for swimming or as gills. Two pairs of antennae. Almost always aquatic and common in marine environments. Respiration by gills or through the general body surface.
Class	MYRIAPODA (centipedes, millipedes, symphilids) No separation between thorax and abdomen. Many pairs of legs, with one or two pairs on each body segment. One pair of antennae. Tracheal system present. Terrestrial.
Class	ARACHNIDA (spiders, scorpions, harvestmen, mites, ticks) Head and thorax merged. (In the mites and ticks all three body regions merged into one single unit.) Four pairs of legs. No antennae. Tracheal system usually present. Mostly terrestrial but some aquatic forms.
Class	INSECTA (insects) Body divided into head, thorax and abdomen. Three pairs of legs. Single pair of antennae. Tracheal system present (except in lower forms). Mostly terrestrial, but some groups common in fresh water. Never fully marine, but some species are intertidal.

Fig. 1. The relationship between insects and other major groups in the animal kingdom

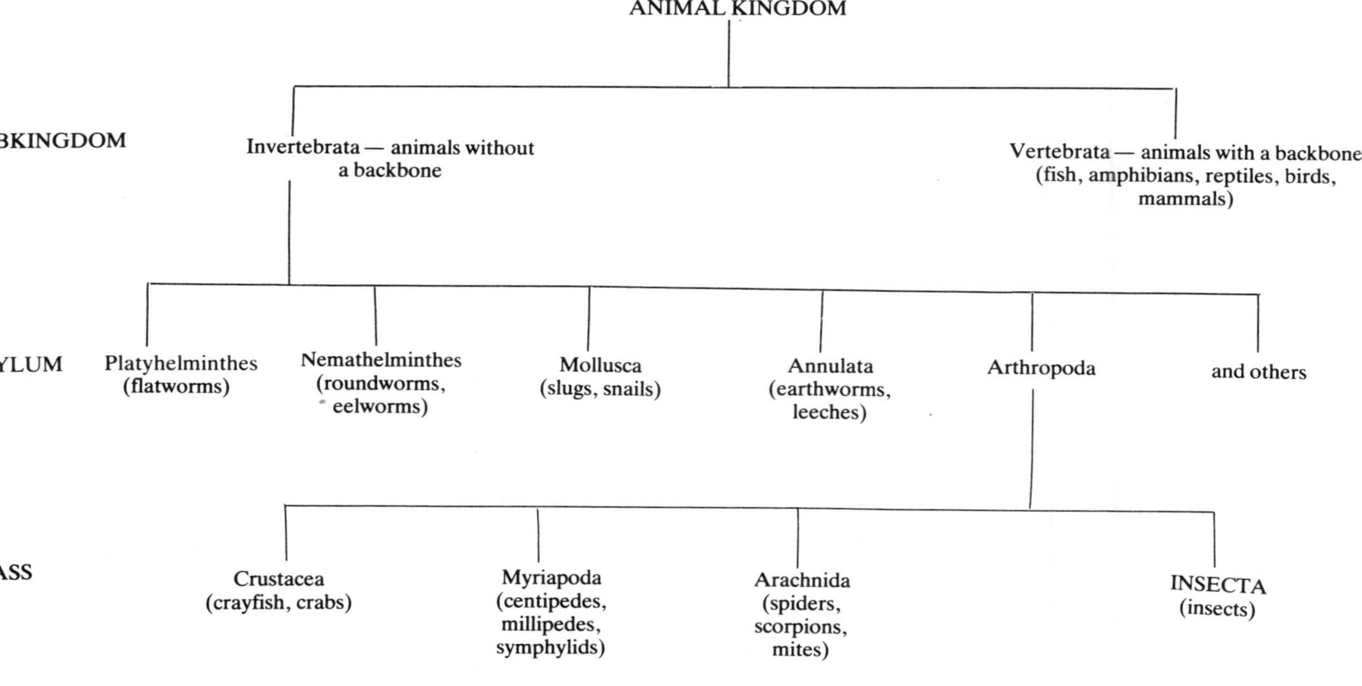

(a) *Large number of species*

Arthropods comprise some 85 percent of all living animals and of these the vast majority are insects, of which there are more than a million different species. We do not in fact know exactly how many exist as entomologists have not yet identified them all. What we can say with certainty however is that there are more species of insects than all other animals put together. Great Britain has about 20,000 species, Australia about 55,000 and New Zealand about 10,000, which are rather small percentages of the world total.

(b) *Large numbers of individuals*

Besides the innumerable different types of insects which exist, individual species often occur in vast numbers. Within a single hectare of grassland for instance there may be several million beetle larvae or grasshoppers, while smaller insects such as aphids and springtails occur in even greater numbers. In tropical countries, swarms of locusts have been recorded more than a hundred kilometres long and dense enough to darken the sky. Such swarms may include as many as 10^9 individuals (10 followed by 8 zeros). Insects individually may be quite insignificant but collectively their impact can be catastrophic.

(c) *Great variety of habitats*

Although particular species of insects may have rather special requirements of food and environmental conditions which limit their distribution, as a group they occur in a great variety of habitats and are found almost everywhere that life can exist, other than in the sea. As they are very temperature dependent they flourish best (both numerically and in variety), in the tropics and thrive less well in temperate climates. Insects still occur in limited variety, though sometimes in great numbers, in sub-arctic regions.

(d) *Long geological history*

Insects do not preserve well as fossils so that our knowledge of their geological history is sparse, but we do know that they have existed on earth for a very long period of time, at least from Carboniferous times some 300 million years ago. During this vast period of time many major groups of animals have come and gone from the scene, but the insects have remained and thrived. Indeed, some groups of present-day insects seem to have changed hardly at all since this early epoch (eg, cockroaches), surely an indication of a very successful "design". Compared to man's puny 2-3 million years of existence the record of insects is formidable. Some scientists even predict that insects will continue to exist long after man has rendered the earth uninhabitable for his own species. Their track record contains nothing to deny such an assertion.

Some possible reasons for insects' success

While no one can be certain why insects have thrived so well, there are a number of factors that may well have contributed:

(a) *Power of flight*

Besides the insects only birds and bats among present day animals have succeeded in developing the power of flight. Flight imparts great mobility to an organism and enables it quickly to colonise new sources of food as they become available (eg, annual crops). Conversely, flight facilitates rapid escape from unfavourable conditions.

(b) *Adaptability*

Insects as a group have adapted to all environments capable of supporting life (other than marine) and moreover can utilise almost any organic material as food.

(c) *Possession of an external skeleton*

Although a handicap in some respects (such as not allowing for growth), an external skeleton provides a small animal with a valuable protective casing. What is perhaps more important than its mechanical strength is the physical property of its outermost layer (**cuticle**) in providing a very effective barrier against water loss, a constant problem for any small land animal.

(d) *Small size*

There are disadvantages in being small, in particular a limited brain size and thus limited capacity for the development of intelligence. For insects, however, the low food requirement per individual and ease of concealment consequent on small size appear to outweigh such a disadvantage.

(e) *Rapid reproduction*

The ability of insects to multiply rapidly (due to short life cycles and many offspring per female) is probably a key factor in their success. Food resources can be quickly exploited as they become available, and furthermore there is the capacity for rapid evolutionary change as shown dramatically by the development of insecticide resistant strains in many species.

Conclusion

Insects as pests present us with a formidable adversary, making up for what they lack in intelligence by sheer weight of numbers and biological versatility. They are unlikely to give up easily in the struggle for existence.

Chapter 2

The Practical Importance of Insects

Man and insects

The ways in which insects affect man's welfare are many and varied. Those insects that are in some way harmful we refer to as pests, but important as these are, they in fact make up only a small minority of insect species. Of the one million or so different species of insects that exist, only a few thousand qualify for pest status. In contrast, many insects are distinctly beneficial to man in that they act as natural enemies of harmful species, as pollinators of many cultivated plants, or as producers of valued materials such as honey and silk. The bulk of insect species, however, fall into neither of these categories of clearly harmful or beneficial, but nevertheless are extremely important as essential components of both natural and modified ecosystems. The main ways in which insect activities affect man's welfare are set out in Table 2.

Harmful insect activities

Although the number of species of insects that are regarded as pests is not large, their activities have major impact on human welfare. The Food and Agriculture Organisation of the United Nations has estimated that *one third* of all food grown is lost to pests and diseases, either from the growing crop in the field or in store after harvest. Although this figure includes losses from disease organisms as well as pests it may be safely claimed that at least half these losses are due to insects. In a world in which many people still go hungry, insects thus still claim more than their fair share.

To some extent such estimates are inflated by the high incidence of insect activity in the tropics and by inadequate pest control programmes in many developing countries. However, pest losses are not by any means limited to such situations. In the United States of America with its highly developed agriculture and pest control technology, the US Department of Agriculture estimated that losses from insect pests totalled US $6.8 billion *annually* in the

Table 2. Insect activities in relation to man's welfare

Value of activity relative to man	*Type of activity*	*Effects*
HARMFUL	Pests* of cultivated plants.	Reduction in yield and quality of produce. Transmission of plant disease.
	Pests of farm animals.	Discomfort. Reduction in vigour and growth rate. Lowered production (dairy stock). Damaged skins and hides. Transmission of disease.
	Pests of stored produce (principally human and animal foods).	Accelerated deterioration. Reduction in quality and nutritional value. Aesthetically offensive.
	Pests of timber and wood products.	Accelerated deterioration resulting eventually in structural failure.
	Pests of medical and public health importance.	Discomfort. Ill health. Reduction in vigour. Transmission of disease.
	Pests in households and industrial premises.	Aesthetically offensive. Damaging to stored food and other produce, eg, wool and wollen goods. Unhygienic.
BENEFICIAL	Natural enemies of pest species (may include undesirable plants, eg, weeds, as well as pest insects).	Suppression of pest or potential pest.
	Pollinators of cultivated plants.	Provide essential pollination of many cultivated plants,especially many horticultural crops.
	Producers of useful materials.	Production particularly of honey and silk.
NEUTRAL	Components of natural and modified ecosystems.	Important components of biological systems essential to man's long-term welfare.

**Note*: It is the *type of association* of an insect with man that justifies the label pest rather than the species itself, eg, honey bees are in the main highly beneficial but may be regarded as pests by growers of glasshouse cucumbers who do not want their plants pollinated as this results in deformed fruit.

decade 1950–1960, including the cost each year of control measures (principally chemical) of US $730 million. The situation is no doubt similar in many other parts of the world.

Some plant feeding insects are important, not so much because they directly reduce plant yield and quality, but because they transmit disease organisms. This is particularly the case with plant virus diseases many of which can only spread from plant to plant by means of suitable insect vectors. In temperate climates aphids of various kinds are the most important vectors while in the tropics leafhoppers are the main culprits. As there is at present no practical cure for a virus infected plant, control is usually aimed at preventing infection through control of the insect vector. Detailed consideration of the ways in which insects injure plants, how plants respond to such injury, and how this in turn is reflected in lowered yield and quality of produce, is included in Chapter 8 of this book.

Livestock farmers have to contend with insect pests directly affecting their animals in addition to those that are plant feeding. Such pests as biting flies, blow flies and lice cause major losses in animal production throughout the world each year, and so do those close relatives of insects, the ticks. The effects of such pests include reduced growth rate, lowered productivity and even mortality in some cases. At slaughter, skins and hides may be of inferior quality. In addition, as with plants, insects may be important indirectly as carriers of disease organisms.

Besides these impacts of insect pests in agriculture and horticulture there are several other ways in which insect activities are harmful to man's welfare, as indicated in Table 2, such as their ability to attack stored food and other products. In addition, the importance of their role as vectors of many major human diseases in tropical countries (such as malaria and yellow fever) can hardly be overstated. This book is about insects in agriculture and horticulture and does not include detailed consideration of their importance in other situations, but the principles of pest control are the same whatever the circumstances.

Beneficial insect activities

There are three main ways in which insects are beneficial from the human point of view; as natural enemies of pest species, as pollinators of cultivated plants and as producers of useful materials such as silk and honey.

(a) *Natural enemies of pest species*

Although insects, like other organisms, are limited by environmental factors such as climate and disease, often the main regulators of populations are other insects which attack and feed on them as parasites and predators. Almost all insects are affected in this way to some extent, some much more than others, so that without natural insect enemies most pest problems would be much more severe and many species that are now unimportant would be raised to pest status. A detailed account of these natural enemies is given in Chapter 9. Their utilisation to help suppress pest species is the basis of biological control and modern pest management and is discussed in Chapters 11 and 12.

(b) *Pollinators of cultivated plants*

Many cultivated plants are dependent on insects for pollination and effective crop production. This applies to virtually all fruits and also to many of those plants which are grown for their seed. Many other crops also require insect pollination for seed production. Naturally occurring insect populations bring about some degree of pollination but often this is quite inadequate for practical purposes. In such cases it is necessary to supplement natural pollination with colonies of honeybees brought in specifically for that purpose. Some plants, however, are not easily pollinated by honeybees, eg, lucerne, and other species of bees may then be managed to improve seed yields. The topic of insect pollination is discussed further in Chapter 8.

(c) *Producers of useful materials*

Honeybees have been managed by man for many hundreds of years and apiculture was well known to the Romans. The prime interest in such management of bees was for honey (which has been used as a sweetener long before sugar) and to a lesser extent beeswax and other minor products. Silkworm culture, which originated in the Far East, has even earlier historical roots and natural silk (from the cocoon of the silkworm moth) has been prized down the ages for its beauty and as a symbol of wealth. It is still not fully matched by the best synthetics and therefore production of natural silk continues. The silkworm is the only truly domesticated species of insect, and is now dependent on man for its continued existence, at least in its cultured form.

In the past, many other insect derived products have been important in human economy, such as shellac and cochineal (both derived from scale insects). In most cases these products have now been superseded by synthetic materials.

One further useful role of insects which is not greatly to Western taste, is their use as human food to supplement or enliven otherwise poor and restricted diets. Insects are still eaten by various peoples in some parts of the world, and not entirely in primitive societies. Nutritionally they can provide a valuable addition to many diets and could be so utilised more widely if it were not for our illogically based aversions.

Neutral insect activities

Most insect species are in fact neither directly beneficial nor directly harmful to man, but this does not mean to say that they are not important in the total realm of nature. Some of these "neutral" species are for instance amongst the early colonisers of dead plant and animal remains assisting the processes of decay and providing for the recycling of the chemical building blocks from which living organisms are formed. Others form major links in the food chains of many small mammals, birds and freshwater fish. What insects lack individually in bulk they often make up for in numbers so that their total biomass may exceed that of any other animal group in many natural systems.

Our knowledge of the detailed functioning of natural (and modified) ecosystems is far from complete so that we cannot define precisely the role that "neutral" insect species fulfil. However, because of their huge variety and large numbers we would be unwise to underestimate the importance of insects in this respect. As man's long-term existence may well depend on the viability of the world's major ecosystems, they may in fact be much more important than we at present imagine.

SELECTED REFERENCES

Atkins, M. D. 1978. *Insects in perspective*. McMillan, New York. 513 pp.

Dethier, V. G. 1976. *Man's plague. Insects and agriculture*. Darwin Press, Princeton. 237pp.

Chapter 3

Insect Structure and Function

In this Chapter the main features of insect anatomy (both external and internal) are described, and, where appropriate, a brief account given of their functioning. A mature cricket, grasshopper or cockroach provides a suitable specimen for examination as it displays most body structures in a fairly generalised form. Details of anatomy vary widely in different insect groups so that the following account is necessarily of a rather general nature. Although the way in which the various body parts function is fairly well understood, it is not possible to ascribe reasons to every modification of form.

External anatomy

The outer body surface of insects, as of all arthropods, consists of a hardened layer which acts as a skeleton for bodily support and as a protective covering to the soft internal organs. It is referred to as the **exoskeleton**. In most insects it consists of a series of hardened plates with flexible membranous areas in between to allow for movement, except in the head where the plates are fused together to form a rigid hollow sphere, the **head capsule**. The general arrangement is thus very much like a suit of armour. In some cases, eg, caterpillars, most of the exoskeleton, except for the head and legs, is soft and flexible whilst in a few insects, such as fly maggots, there are no hard parts at all.

In all cases, however, whether the external skeleton be hard or soft, the outer surface consists of a waxy layer (**cuticle**) which is highly impermeable to water. This outer layer is extremely important to insects in minimising loss of water from the body and its occurrence is one of the reasons insects have been so successful as land animals despite their small size. If the cuticle is damaged in any way, by an abrasive dust for instance, the insect may lose moisture rapidly and die.

(a) *Segmentation*

As pointed out in Chapter 1, insects are segmented creatures, the body consisting of a series of similar units or **segments**. Individual segments can be seen most readily in the hind end (**abdomen**) which often presents a horizontally striped appearance. In other parts of the body (the head in particular), the segments tend to be more modified and merged together. Internally the segmented pattern is clearly apparent in the nerve cord and respiratory system (see later sections of this Chapter).

The body segments of insects are grouped into three body regions (**head, thorax** and **abdomen** from front to rear) which may be very distinct and separated by constrictions, as in bees and wasps; or less so as in caterpillars. These body regions should not be confused with individual segments. There are a number of segments to each body region as indicated below. The main features of the external anatomy of a typical insect are shown in Fig 2.

(b) *The head*

The head consists of at least six segments but no direct evidence of their presence can be seen as they are fused into a single structure, the **head capsule** which is equivalent to the skull in vertebrates. Inserted into the head capsule are the main sense organs of insects, the eyes and the **antennae**. (Sense organs are discussed more fully later in this Chapter.) The lower part of the head is occupied by a group of structures which comprise the **mouthparts**. Their form varies greatly in different insect groups, in each case being modified according to the nature of the food and mode of feeding. A detailed consideration of insect mouthpart structures follows in the next section.

(c) *The thorax*

The middle body region of insects is the thorax, and always consists of three segments. These may be more or less distinct but often their margins are difficult to identify as the overall structure of the thorax is rather complex to allow for the attachment and operation of legs and wings. Insects typically possess three pairs of legs, one pair to each segment of the thorax. Each leg consists of four main portions with flexible joints between. At the extremity of the leg is the foot (**tarsus**) which is usually composed of several joints. The tarsus normally terminates in a pair of claws though there may be only a single claw. In addition, pad-like structures to provide extra grip are also often present.

Wings when present arise from the second and third segments of the thorax. There are typically two pairs but in a few groups, such as the true flies (Diptera), they are reduced to a single pair. Many insects are of course wingless. The wings themselves are mostly membranous but in beetles (Coleoptera) the forewings, take the

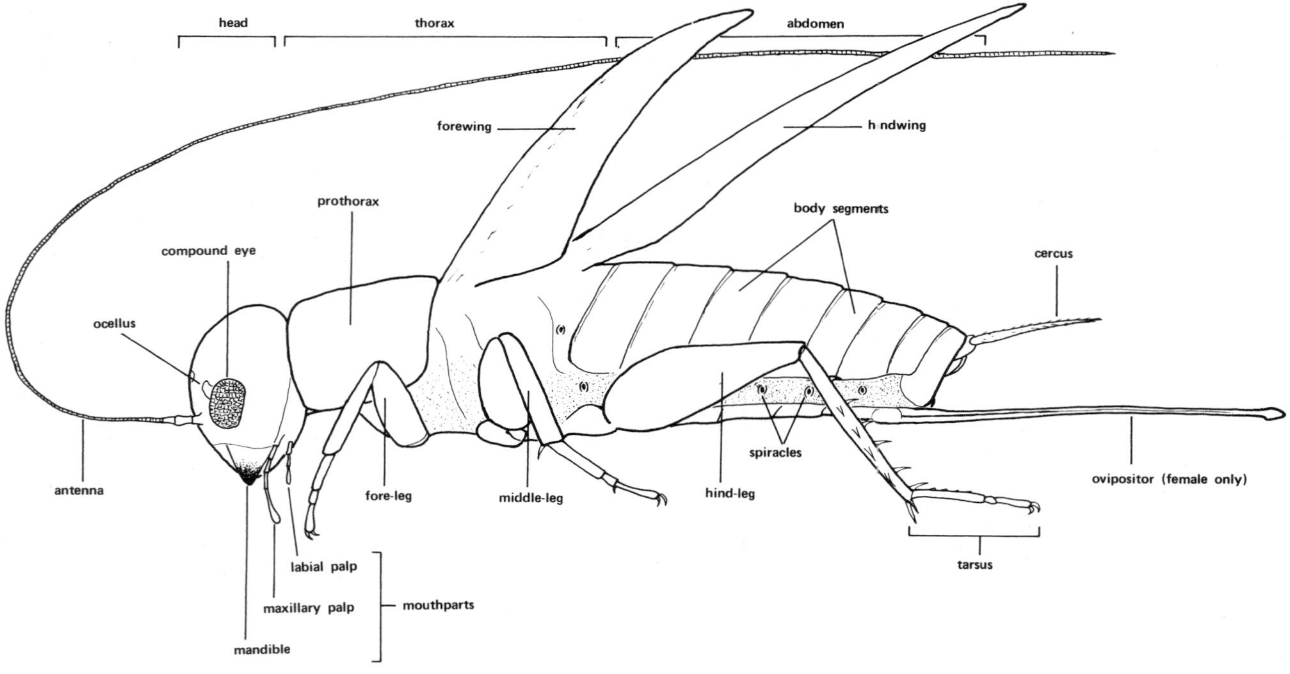

Fig. 2. The main external features of a typical insect (black field cricket).

form of hard horny cases which are not involved in flight. In all except the smallest of insects the membranous part of the wing is supported by a series of thickened ribs, called **veins**. The pattern of arrangement of these veins, their branches and cross connections, varies greatly in different insect groups and is an important feature in classification. A range of insect wing types is shown in Fig 3.

Most insects possess some mechanism for folding the wings when not used for flight. Unlike birds, there are no muscles extending into the wings themselves. Instead they are operated by muscles within the thorax which, when they contract, distorts in shape and thus causes the wings to beat.

In the soft membranous areas between the hardened body plates on the sides of the thorax and abdomen are minute openings, rather circular in outline and dark in colour. These are **spiracles** which are the external pores of the insect's respiratory (**tracheal**) system. There is typically a pair of spiracles to each body segment but the actual number and arrangement on different segments varies considerably. They often can be seen more readily along the sides of the abdomen rather than on the thorax.

(d) *The abdomen*

The hind body region of an insect is the **abdomen**. In its primitive form it consists of eleven segments though fewer than this can normally be counted as the hind ones tend to be reduced and retracted one inside the other. A row of spiracles commonly occurs along each side of the abdomen, one pair to each segment. In the more primitive insect groups a pair of tail filaments (**cerci**) arise from the tip of the abdomen, and there may be a central third one. In more highly evolved insects they are reduced in size or absent.

External reproductive organs are present on the abdomen of many insects. In the female these may take the form of a sword-like **ovipositor** down which eggs pass at the time of egg laying. It consists of up to three pairs of shafts and arises from the eighth and ninth abdominal segments. It is well developed in crickets and cicadas. In some insect groups such as the true flies (Diptera), the hind end of the abdomen is telescopic in form and acts as an ovipositor instead.

In male insects external reproductive organs involved in copulation can often be seen at the tip of the abdomen, although in many cases they are largely enclosed and not readily visible. The most obvious feature is a pair of **claspers** of varying shape and size with which the abdomen of the female is grasped during mating. A central copulatory organ (**penis, aedeagus**) is located between the claspers.

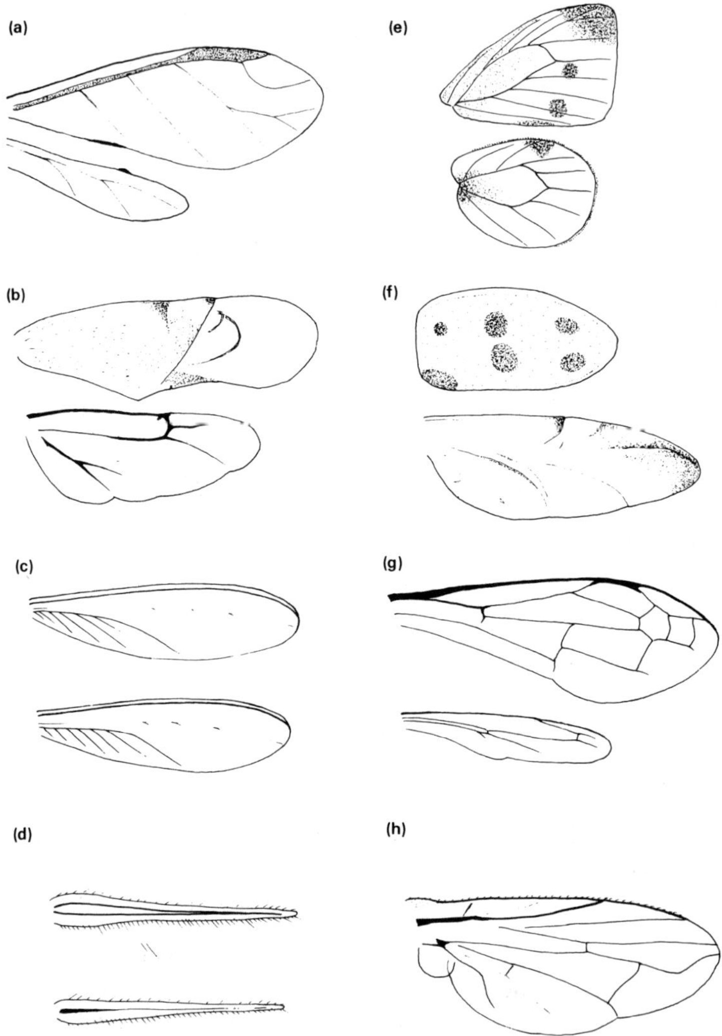

Fig. 3. Some patterns of insect wings;

(a) Aphid (Hemiptera—Homoptera),
(b) Mirid (Hemiptera—Heteroptera),
(c) Termite (Isoptera),
(d) Thrips (Thysanoptera),
(e) Butterfly (Lepidoptera),
(f) Ladybird beetle (Coleoptera),
(g) German wasp (Hymenoptera),
(h) Housefly (Diptera).

Internal anatomy

As insects have to provide for the same bodily functions as higher animals the main body systems that exist in vertebrates also occur in insects. However the nature of these systems in insects is very different from those of higher animals. The main body systems that can be identified are:

skeletal system
circulatory system
digestive system
excretory system
respiratory system
nervous system
reproductive system.

(a) *Skeletal system (skeleton)*

As the outer body casing of insects (exoskeleton) fulfils the main function of a skeleton (that of providing support and protection for soft body parts) there is little internal skeletal development. Provision must be made however for the attachment of muscles and this is achieved by internal protrusions of the exoskeleton in some parts of the body (particularly in the thorax to anchor the large muscles which operate the wings and legs). There is usually some development of an internal skeleton in the head to provide extra strength and rigidity to the head capsule and to give attachment to muscles that operate the jaws of biting insects.

(b) *Circulatory system*

The functions of the circulatory (blood) system of higher animals are to distribute food materials to all parts of the body, to carry away waste products, and to circulate hormones. In addition, as part of the respiratory cycle, blood must also carry oxygen to, and carbon dioxide away from tissues. The pigment haemoglobin imparts these properties and gives vertebrate blood its bright red colour. In insects, however, respiration is provided directly by a system of air tubes (see below) so that insect blood (**haemolymph**) is not required to carry oxygen or carbon dioxide and so is not pigmented.

The internal body cavity of insects (**haemocoel**) is filled with haemolymph which is colourless, or nearly so. In order to provide for the other functions of a circulatory system besides the respiratory one, haemolymph must circulate around the body. This is achieved by the **dorsal blood vessel** which is a muscular tube lying in the haemocoel close to the dorsal (upper) surface. It has a series of inlet valves (**ostia**) along most of its length, typically a pair to each body segment, which take in haemolymph. Contractions of the blood vessel then force the haemolymph forward along its length to discharge in the head region. The fore part of the dorsal blood vessel is simply a tube, and is referred to as the **aorta**. The hind part is muscular and contractile and is the **heart**. In insects with a thin transparent exoskeleton, eg, some caterpillars, regular pulsations of the heart can be seen through the body wall, down the mid-line of the dorsal surface. There are no true blood vessels in insects other than the dorsal aorta so the process of circulation is a fairly crude one. It may be assisted in some cases by perforated membranes which divide the haemocoel into a series of spaces (**sinuses**), and by accessory pumps at the base of the wings and legs. The general plan of the circulatory system of an insect is shown in Fig 4(a).

(c) *Digestive and excretory systems*

The alimentary canal of insects consists of a relatively straight tube running from the mouth on the lower part of the head to the anus at the tip of the abdomen. There may be some side branches and slight convolutions but there are no long coiled portions comparable to the intestines of higher animals. Thus the total length of the gut is usually not much more than the length of the insect itself. Three regions of the gut are recognised — **foregut, midgut** and **hindgut**. These are more than simply different portions of the gut because the layer of cuticle, which covers the external body surface of insects, extends into and lines both the foregut and the hindgut. This lining is shed and renewed along with the general body surface each time the insect moults. The midgut possesses no such lining. The walls of the foregut and hindgut are rendered rather impermeable by this cuticular lining and most digestion and absorption of food therefore takes place in the midgut region. The general layout of the alimentary canal of a typical insect is shown in Fig 4(b).

From the mouth a tubular portion (**oesophagus**) leads back to a distended area (**crop**)* for reception and temporary storage of food. **Salivary glands** which discharge through the mouth are well developed in most insects. The saliva they produce may assist in partial external digestion. Little digestion occurs in the foregut but some grinding of food may take place and some insects, such as cockroaches, have special hardened internal teeth-like structures

* In insect anatomy many terms are "borrowed" from vertebrate anatomy. They usually refer to the same sorts of structures which often have similar functions but it should be remembered that they are not identical and may have arisen in a quite different way in the course of evolution.

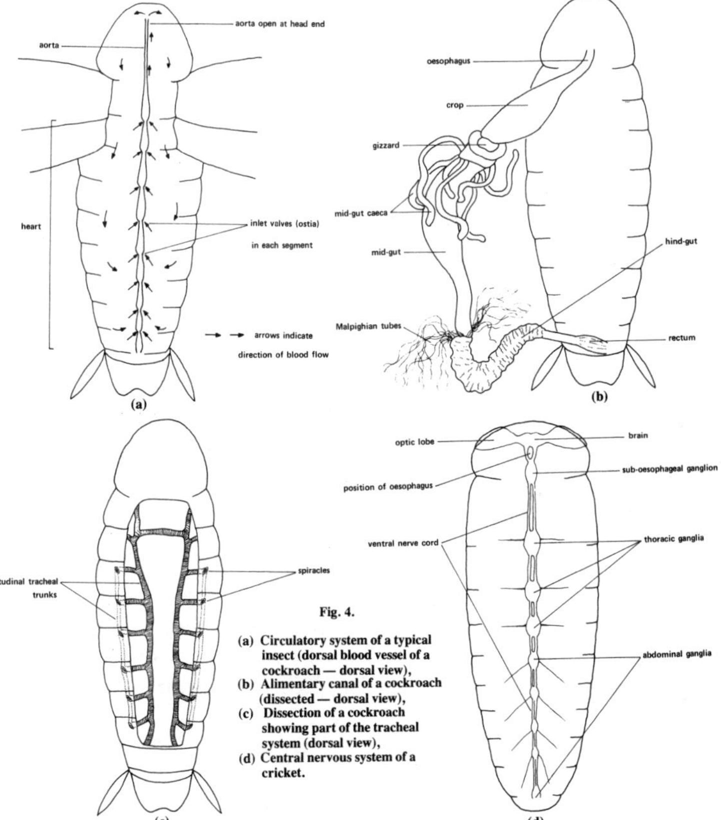

Fig. 4.

(a) Circulatory system of a typical insect (dorsal blood vessel of a cockroach — dorsal view),
(b) Alimentary canal of a cockroach (dissected — dorsal view),
(c) Dissection of a cockroach showing part of the tracheal system (dorsal view),
(d) Central nervous system of a cricket.

for this purpose. Behind the crop there is a muscular sphincter which opens intermittently to allow food through to the midgut region.

The midgut is usually a rather wide and fairly straight tube. From its front part, in most insects, several finger-like outgrowths (**gastric caeca**) arise. These provide additional internal surface area for the digestion and absorption of food.

Insects produce no mucus to lubricate the passage of food along the gut. To provide protection to the delicate midgut inner surface a special membrane (**peritrophic membrane**) is continually secreted from a ring-shaped gland at the beginning of the midgut region. It envelops the food and travels back through the gut with it, slowly disintegrating and being resorbed in the process. It is present in all insects except those that are liquid feeding. A number (1 to more than 100) of fine tubes connects with the hind part of the midgut. These are the **Malpighian tubules** which are excretory in function and thus comparable to the kidneys of higher animals. They lie in the haemolymph from which they collect waste products and discharge into the gut. They are often yellowish or chalky-white in colour.

The hindgut usually shows two distinct portions: a tubular intestinal part in front and a thick muscular **rectum** behind opening in the anus. Considerable absorption of moisture from digested food waste takes place in the hindgut so that in most insects what is finally excreted is fairly dry. In some insects micro-organisms within the gut assist with breakdown and digestion of food. This is particularly the case with cellulose feeders such as termites.

(d) *Respiratory system*

Insects breathe by a system of air tubes (**tracheae**) whereby oxygen is taken direct to body tissues from the outside. There is no interchange of oxygen with the blood (haemolymph) and so no organs comparable to lungs are required. Such a method of respiration is called a **tracheal system** and occurs only in insects and a few close relatives. The **spiracles**, which are the external openings of the system, lead in to a network of air tubes (tracheae) which are lined with a thin layer of cuticle. There are usually several large longitudinal tracheal trunks with cross connections in each body segment. From these main trunks branches arise which divide and subdivide more finely to feed all parts of the body. The very finest branches (**tracheoles**) are microscopic and are filled with fluid rather than air. It is here that gaseous exchange with the tissues takes place. Respiration for most insects is a passive process of diffusion but passage of air through the tracheae may in some cases be assisted by regular movements of the body, as for example in the telescopic movement of the abdomen shown by honey bees. In insects that fly, parts of the tracheal system are distended into a series of internal air sacs. The basic layout of the tracheal system in a cockroach is shown in Fig 4(c).

(e) *Nervous system*

Only the general organisation of the central nervous system will be discussed here. There is also another less obvious nerve network that regulates activity of the alimentary canal and other internal organs.

The basic layout of the central nervous system (see Fig 4(d)) consists of a double nerve cord running along the ventral (lower) side of the insect close to the body wall. There is a swelling (**ganglion**) in each segment from which branches arise to serve adjacent parts of the body. In the head region three ganglia are fused together to form a larger nerve mass immediately below the oesophagus — the **sub-oesophageal ganglion**. A further three ganglia make up another and larger nerve mass in the upper part of the head — the **brain**. The double nerve cord connecting the brain with the sub-oesophageal ganglion separates to allow passage of the oesophagus. The eyes and antennae are connected directly by short nerves to the brain. In some more highly evolved insects, eg, some flies, the longitudinal nerve cord is very abbreviated and the ganglia reduced in number, even to a single thoracic nerve mass.

A feature of the functioning of the nervous system of insects is that various parts of the body are much more under local nervous control (autonomous) compared to higher animals. Perhaps as a consequence much insect behaviour is stereotyped, and there is limited ability to learn from experience and to modify behaviour according to circumstances. Most behavioural patterns, even complex ones such as nest building in social insects, are inherited and thus imprinted into the nervous system from birth.

(*f*) *Reproductive system*

In most insects reproduction depends on mating between the sexes, followed by egg laying. Internal reproductive organs therefore consist of **ovaries** and **testes** in the female and male, respectively, plus associated structures concerned with the acts of mating and egg laying. As the reproductive process of insects is considered in detail in Chapter 5, description of the organs concerned is given there.

(g) *Fat body*

Much of the space around the internal organs of an insect is often packed with irregular shaped white material which may obscure much of the gut and other internal organs. This is **fat body** and is a storage organ by which the insect lays down food reserves at one stage in its development for utilisation at a later stage. It is particularly prominent in many newly emerged adult insects which may not feed at all or only in a limited fashion, and thus depend on energy supplies laid down at an earlier stage in the life cycle.

Insect mouthparts

The injury which insects inflict on plants is almost invariably a result of feeding activity. The few exceptions to this are where injury results from egg laying, as for instance with cicadas. Different insects feed in different ways so that some knowledge of the structure and functioning of their mouthparts is essential for proper understanding of the nature of plant injury in its various forms.

Amongst plant feeding insects two main patterns of mouthparts are recognised — the **biting and chewing pattern** in those that consume solid plant tissue, and the **piercing and sucking pattern** in those that are sap feeders. A third type, usually referred to as the **rasping pattern**, also occurs but is less important. Each of these types will be considered in terms of structure and functioning. In addition to destructive feeding on plants, insects also utilise many other food materials. Specialised adaptations of the mouthparts often occur in such cases and examples of several of these will be referred to briefly.

(a) *Biting and chewing mouthparts*

Besides biting off portions of plant tissue and chewing these into smaller more manageable portions for ingestion, sensory evaluation of food must also take place before feeding proper commences. Although sense organs other than those on the mouthparts may be involved in the evaluation of food, eg, those on the antennae, it is sensory structures on the mouthparts themselves that are primarily concerned, in particular the feeler-like **palps**.

The basic biting and chewing mouthpart pattern is present in crickets and cockroaches (see Fig 5). Details can only be seen if the various parts are carefully dissected free from the head and spread out. There are four main parts. In front, a flap-like front lip (**labrum**) acts as a protective cover to the large biting jaws (**mandibles**) which lie immediately behind it. Minute organs of touch and taste may be present on the rear (inner) surface of the labrum. The mandibles are hinged so that they act one against the other when they meet in the mid-line. Their inner edges are often serrated. In some predatory insects, eg, lacewings, the mandibles are

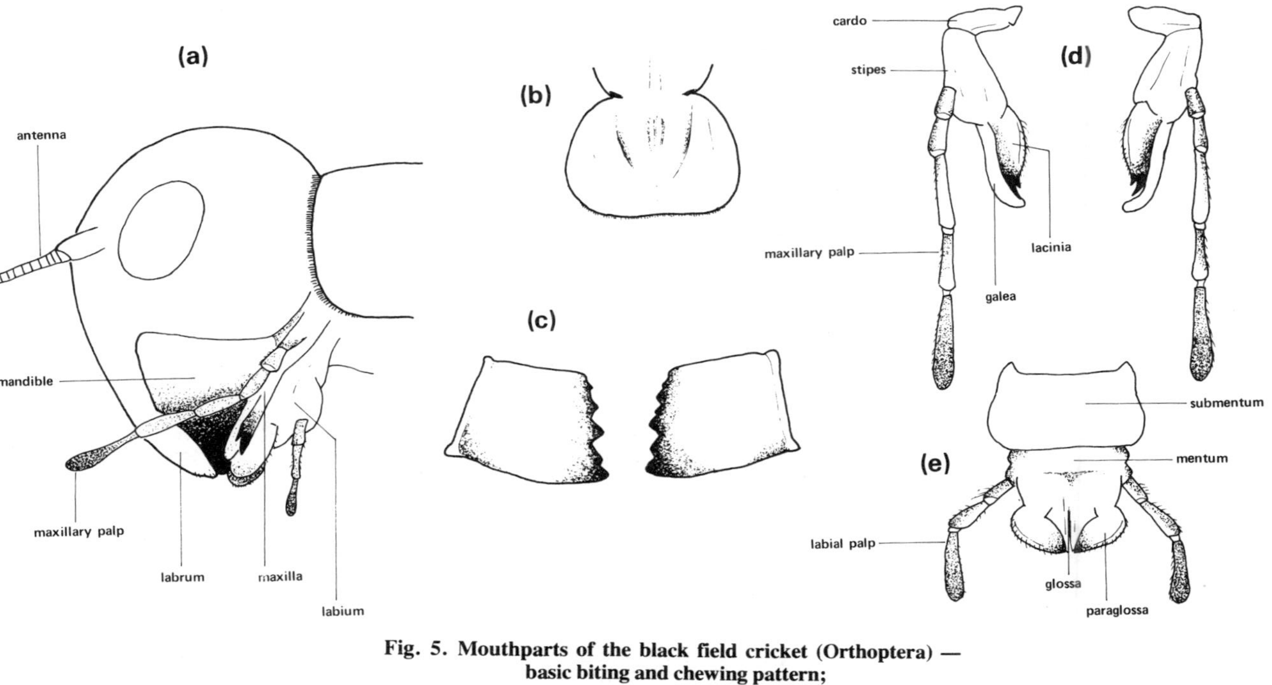

Fig. 5. Mouthparts of the black field cricket (Orthoptera) — basic biting and chewing pattern;

(a) head in side view showing mouthparts ***in situ*****,**
(b) labrum, (c) mandibles, (d) maxillae, (e) labium.
(b), (c), (d) and (e) dissected free from head.

long and pointed for piercing their prey, rather than for chewing. The mandibles reach enormous proportions in some wood chewing insects.

Behind the mandibles and slightly to each side are a pair of supplementary jaws (**maxillae**). Each maxilla consists of a basal portion made up of two joints, a pair of pincers of varying shape, and a feeler-like palp borne as a side branch. The maxillae are both manipulatory and sensory in function, the palps being clothed in minute organs of taste and touch. Finally, behind the maxillae in the form of a hind lip is the **labium**. This is a single structure but has the appearance of two small maxillae linked together and sharing a common base of attachment. As with the maxillae there are both manipulatory and sensory parts, the latter in the form of a second pair of palps somewhat smaller than those on the maxillae.

The true mouth of insects (ie, the opening to the gut), is situated between the mandibles and maxillae. Adjacent to it there is often a fleshy lobe called the **hypopharynx** which acts as a tongue.

Considerable departure from this basic pattern occurs in various insect groups which nevertheless retain the biting/chewing type feeding behaviour. This is particularly the case with larvae (caterpillars) of butterflies and moths (Lepidoptera), as shown in Fig 8A. The form of the labrum and mandibles are much as in the cricket or cockroach, but the maxillae and labium are merged into a single structure and the palps of each reduced to small peg-like outgrowths. Their sensory function is however retained. In many beetles (Coleoptera) the labium is reduced to a small plate with tiny palps, though the maxillae remain well developed.

(b) *Piercing and sucking mouthparts*

Mouthparts adapted for piercing plants and sucking out the sap are found primarily in the insect order Hemiptera which includes cicadas, leaf hoppers, plant bugs, aphids and mealy bugs. The structure of the mouthparts is much the same in all cases though their size and relative ability to penetrate plant surfaces varies greatly. A cicada* provides a convenient specimen for examination as the various parts can be seen clearly without much magnification. The external appearance and diagrammatic cross sections of such mouthparts are shown in Fig 6.

From the lower part of the head there is a long finger-like extension which at rest tends to be folded back along the underside of the insect. This is the **proboscis** or **rostrum** and is formed from the hind lip (labium). It is tubular in form with a groove-like opening along its lower (front) margin. Within the groove lie two pairs of hard but

*Cicadas will not be readily available to British students. A plant bug (family Miridae or similar) though considerably smaller, provides an adequate substitute.

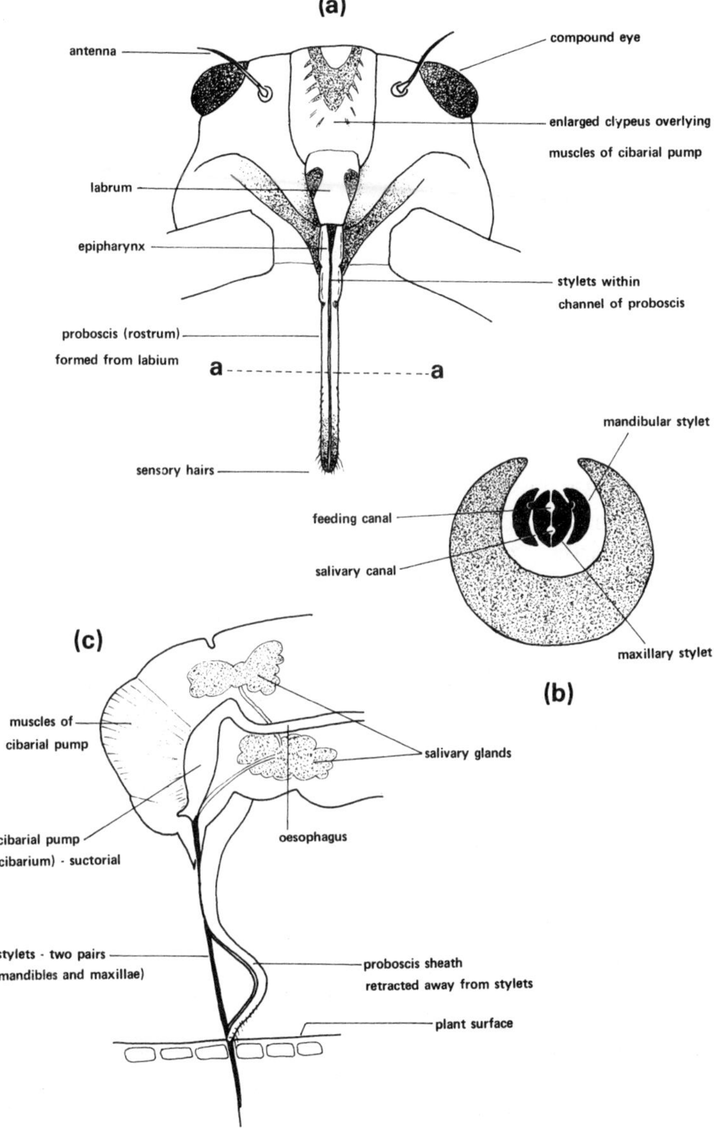

Fig. 6. Mouthparts of a cicada (Hemiptera) — piercing and sucking pattern;

(a) general view of head from beneath,
(b) diagrammatic cross section of proboscis, a------a,
(c) diagrammatic vertical section of head with mouthparts in feeding position (side view).

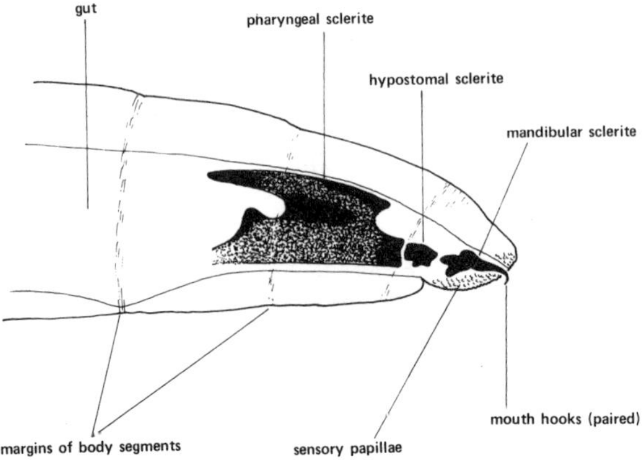

Fig. 7. Mouthparts of a housefly larva (Diptera) — rasping pattern.

flexible bristles. These are the **stylets** and are formed from the mandibles and maxillae. When the insect feeds they are extended from the tip of the rostrum and penetrate the plant. Normally the inner two stylets (formed from the maxillae) are firmly locked together and their inner surfaces so shaped that they form two canals (see Fig 6). One canal allows the insect to pump saliva into the plant; the other, the feeding canal, is for uptake of plant sap. Within the head of the insect the first portion of the gut (**pharynx**) is distended into a sucking pump (**cibarium**) which assists in the feeding process. No palps are present and the labrum is reduced to a tiny triangular flap. Hairs towards the tip of the rostrum act in a sensory fashion.

(c) *Rasping mouthparts*

The third pattern of mouthparts found amongst plant feeding insects is that which occurs in the larvae (maggots) of higher flies (Diptera). It can be seen in the larva of a house fly (Fig 7) though feeding in this case does not take place on living plant material.

The entire insect is soft bodied except for some dark portions embedded in the fore part of the gut and visible through the semi-transparent body. The tips of these hard parts protrude from the mouth in the form of a pair of **mouth hooks**. These are used to shred food material which is then ingested in a semi-liquid state. It is uncertain how mouthparts of this pattern relate to those of other insects and it is likely that they have evolved separately. No palps are present but a number of minute sensory papillae surround the

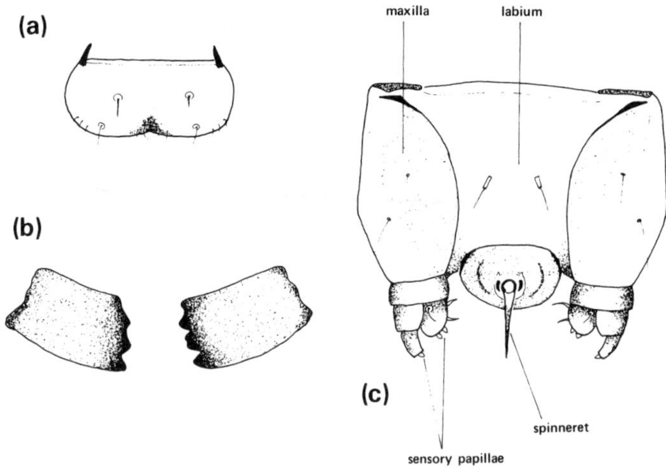

Fig. 8. A. Mouthparts of a moth caterpillar (Lepidoptera)—modified biting and chewing pattern. (a) labrum, (b) mandibles, (c) maxillae and labium.

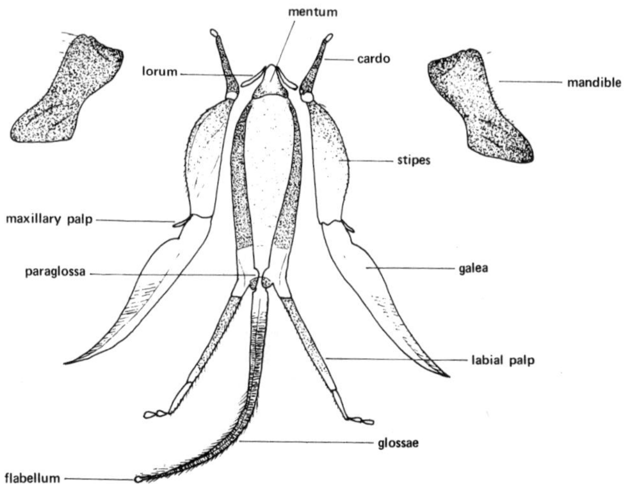

Fig. 8. B. Mouthparts of a honey bee worker (Hymenoptera)—lapping pattern.

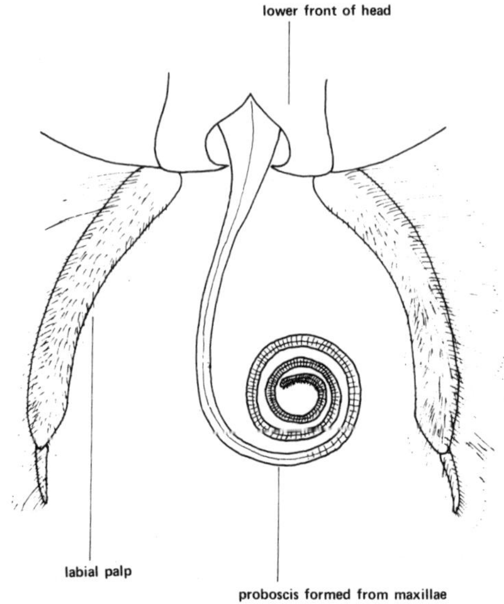

Fig. 8. C. Mouthparts of an adult moth (Lepidoptera)—siphoning pattern.

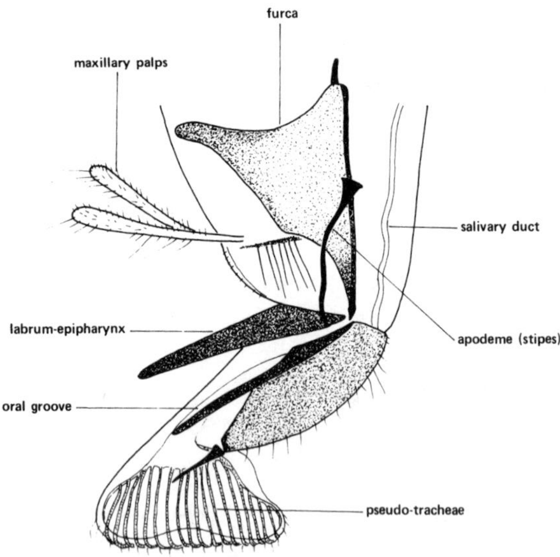

Fig. 8. D. Mouthparts of an adult blowfly (Diptera) — sponging pattern.

mouth. This mouthpart pattern occurs in the plant feeding larvae of various true flies (Diptera) such as carrot fly and narcissus flies.

(d) *Other mouthpart patterns*

In addition to the above, many other variations of mouthpart structure occur in different insects in association with their mode of feeding. Three examples of these in common insects will be considered.

(1) *Adult honey bee (worker) — Lapping pattern* — A major function of the mouthparts of a honey bee is the collection of nectar from flowers. As nectar often lies at the bottom of a rather deep corolla tube the central part of the bee's mouthparts (formed from the inner portions of the labium) is drawn out into a long slender hairy tongue (Fig 8B). Mandibles are present but are used to manipulate wax rather than for biting and chewing and are therefore of slender form.

(2) *Adult moth or butterfly – Siphoning pattern* – Moths and butterflies, like honey bees, feed primarily on nectar but have adopted a different approach to the problem of obtaining it from flowers. Their mouthparts are in the form of a long narrow tube (formed from the maxillae) open only at the tip. This is usually coiled like a watch spring at rest but can be extended and inserted into a flower opening (Fig 8C). It acts like a drinking straw up which nectar can be drawn. A pair of palps (labial) are the only other parts present.

(3) *Adult house fly or blow fly – Sponging pattern* – On the lower part of the head of most flies, including house flies and blow flies, there is an elbowed arm which can be extended. At its tip is a soft pad-like "foot" which is traversed by a series of fine tubes connecting up to the oesophagus. The foot acts like a sponge enabling the fly to mop up liquid or semi-liquid food. A pair of palps (maxillary) are present on the front of the mouthpart arm. The arrangement of parts is shown in Fig 8D.

Note: Only the major patterns of insect mouthparts have been described here. Other variations occur in various groups according to their particular feeding behaviour.

Insect sense organs

Insects possess the five familiar senses of:

touch
hearing
smell
taste
sight

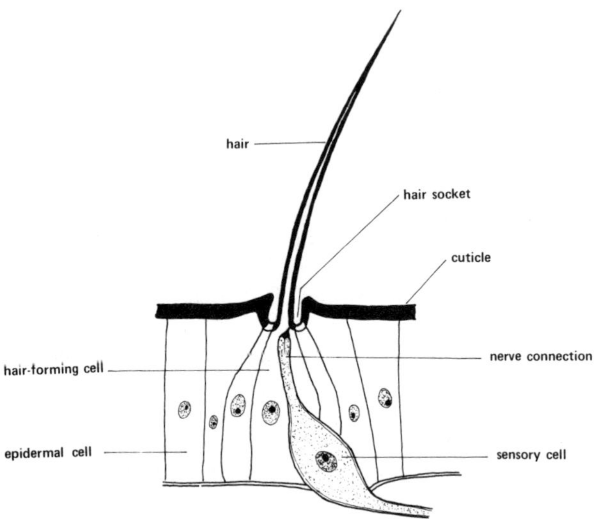

Fig. 9.A. Structure of an insect touch sensitive hair (sensillum).

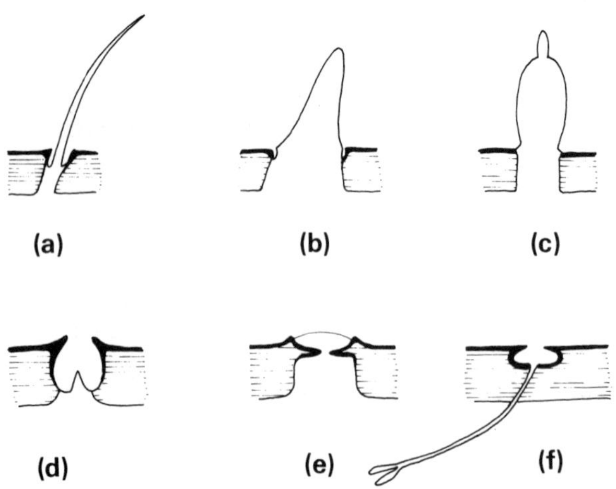

Fig. 9. B. Forms of insect sensory sensilla;
(a) trichoid, (b) basiconic, (c) styloconic,
(d) coeloconic, (e) placoid, (f) ampullaceous.
(after Imms).

However, the sense organs concerned are very different from our own, and so is the relative importance of the various senses in insect biology. For instance, hearing in insects is generally unimportant whereas the chemical senses (of smell and taste) are often dominant.

The basic sense receptor in insects, on which all senses other than vision are based, is an innervated body hair (**sensillum**). With the sense of touch such receptors are relatively unchanged from the basic hair-like form, but with the sense of smell and taste they are highly modified to minute pegs, plates or pits.

(a) *Touch*

The sense of touch in humans is provided by numerous nerve endings just beneath the skin. With insects such a system would not be very effective due to their largely unyielding external skeleton. Instead, many hairs which occur on various parts of the body are provided with a flexible base of attachment and a nerve connection (Fig 9A). When insects brush against something the hairs move in their sockets and trigger nerve impulses. Although most body hairs of insects may be sensitive in this way dense concentrations of minute touch sensitive hairs are often present on the antennae, mouthpart palps and tail cerci.

(b) *Hearing*

The sense of hearing is not important for most insects and there may be no response even to very loud noises. Some insects however use sound as a sex attractant mechanism, eg, crickets, cicadas, and in such cases well developed sound producing and sound receiving organs are present. The latter consist of a membrane of some sort which is set in vibration by sound impulses. Specifically modified groups of innervated hairs associated with the membrane act as detectors.

(c) *Smell and taste*

Both these senses involve the detection of chemical substances in the environment. Smell acts primarily at a distance with chemicals in the vapour phase, whereas taste is by contact only so that chemicals must be in solution. In many natural situations however (an insect biting into a plant for example), both senses are involved and they may be considered together under the term **chemoreception**.

Chemicals play a major role in regulating much insect behaviour and specific substances (or mixtures) may be involved in such processes as selection of food plants, oviposition sites and location of the opposite sex. Chemical senses are thus well developed in

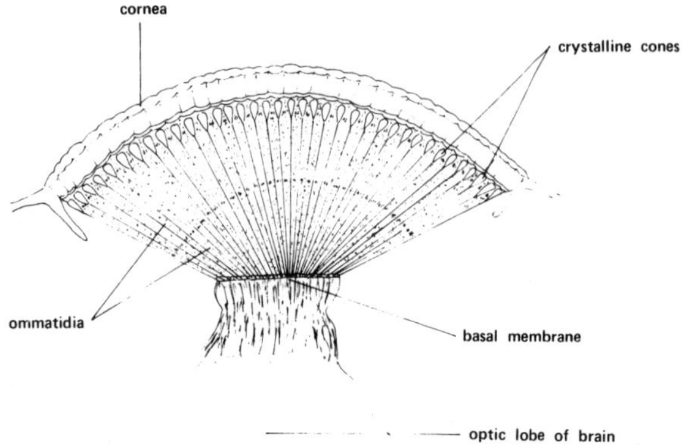

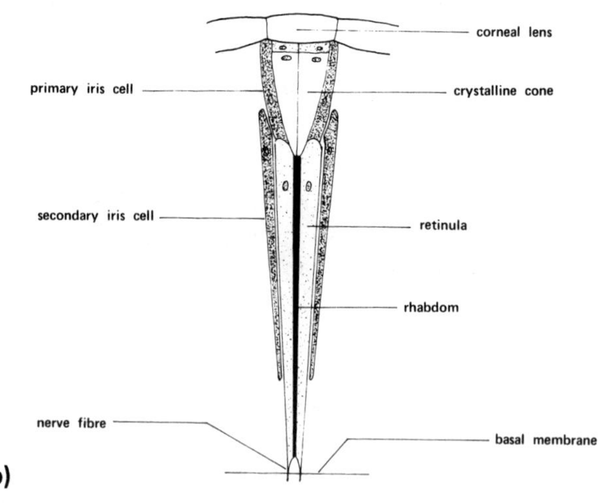

Fig. 10. (a) Section through the eye and optic lobe of a honey bee worker showing the general structure of an insect compound eye.
(b) Diagram of a single unit (ommatidium) from an insect compound eye. (both redrawn from Imms).

most insect species. The basic components of the chemical sense organs are sensilla in the form of minute pegs, plates or pits (Fig 9B). They may be located on many parts of the body, but tend to be concentrated on the antennae and mouthpart palps but can also be present on the tarsi (feet) or ovipositor so that some insects can actually taste with their feet! Individual receptors may be quite specific as to the substances to which they react, though others are rather broad in their response. They are often extremely sensitive. Some butterflies for instance can detect sugar (sucrose) at less than 1/200th the minimum concentration that the human tongue can taste. The ability of many male moths to respond to minute concentrations of the female sex attractant for the species is quite phenomenal, as they react to literally a few molecules. (Insect sex attractants are discussed further in Chapter 5.) The exact mechanism by which chemical molecules impinging on the sensilla generate nerve impulses is unknown, as also are mechanisms which impart specificity.

(d) *Sight*

Insect eyes are of two types: small single lens eyes called **simple eyes** or **ocelli** and large prominent eyes made up of many small units (**compound eyes**). Some insects, particularly those that live in darkness (eg, are soil dwelling) are blind.

(1) *Simple eyes (ocelli)* — Simple eyes consist of a group of light sensitive cells beneath a single semi-transparent lens formed from a specialised area of cuticle. The lens is often in the form of a small raised dome set into the head but may be simply a pale area flush with the surface. Ocelli often occur in groups of up to seven or eight on either side of the head, as in caterpillars, but there may be only three in a triangle on top of the head, as in adult flies (Diptera). Vision provided by ocelli is probably very crude and they may be able to do no more than distinguish between dark and light. Ocelli are the only eyes possessed by insect larvae, though in adult insects they are often present in addition to compound eyes.

(2) *Compound eyes* — Compound eyes, as their names suggests, consist of a number of units combined into a single structure (see Fig 10(a)). Each unit (**ommatidium**) is rather like a simple eye but on a much smaller scale (Fig 10(b)). The number of ommatidia in each eye varies from a few hundred in insects such as ants which live mostly in darkness, to more than 20,000 in adult dragon flies which depend on sight for hunting other insects on the wing.

Each ommatidium is more or less separated from its neighbours by pigment cells, so that what the insect sees is probably a sort of mosaic picture. There is no focussing mechanism and the image cannot be very clear, but detection of movement may be very effective when an object moves across the field of vision and affects different ommatidia in turn. Some insects undoubtedly possess colour vision to some degree and this is particularly well developed in bees which use colours as well as scents to locate flowers. In the case of honey bees vision extends into the ultra-violet range so that their perception of flower colours may be quite different from our own. Honey bees also have the remarkable ability to detect the degree of polarisation of light and to use this as a compass bearing even on cloudy days.

Other specialised sense organs, such as humidity sensors, may be present in some insects but little is known about them.

SELECTED REFERENCES

Introductory

Richards, O. W.; Davies, R. G. 1978. *Imm's outlines of entomology*. 6th edition. Chapman and Hall, London. 254 pp.

Wigglesworth, V. B. 1974. *Insect physiology*, 7th edition. Chapman and Hall. London. 166 pp.

Advanced

Borror, D. J.; De Long, D. M. and Triplehorn, C. A. 1981. *An introduction to the study of insects*. 5th ed. Holt, Rinehart and Winston, New York. 812 pp.

Chapman, R. F. 1982. *The insects: structure and function*. 3rd edition. Hodder and Stoughton/Harvard University Press, 832 pp.

Mordue, W. 1980. *Insect physiology*. Blackwell Scientific, Oxford. 108 pp.

Richards, O. W.; Davies, R. G. 1977. *Imm's general textbook of entomology*. 10th edition. Vol. I. Structure, physiology and development. Chapman and Hall, London. 418 pp.

Romoser, W. S. 1973. *The science of entomology*. MacMillan, New York. 449 pp.

Ross. H. H.; Ross. C. A. and Ross, J. R. P. 1982. *A textbook of entomology*. 4th edition. John Wiley & Sons, New York. 666 pp.

Wigglesworth, V. B. 1972. *The principles of insect physiology*. 7th edition. Chapman and Hall, London. 827 pp.

Chapter 4

Growth, Development, Metamorphosis*

Growth and development of insects show a number of unusual features associated with their biology and mode of life.

Amount of growth

All insects undergo a great deal of growth between the time they hatch from the egg until they are fully grown. Three examples are given in Table 3(a). Even the lowest growth ratio cited is x 450 from egg hatch to mature larva. In the other cases it is much more. The corresponding value for humans from birth to adulthood is only about x 20, but of course for the early part of their development mammals are nurtured within the womb whereas insects are not (except for rare exceptions). The reason for these high growth ratios in insects is not hard to find. To ensure survival most insects must produce large numbers of offspring. Physiologically this is only possible if the eggs are small, which in turn means that a great deal of growth has to take place before adulthood is again reached.

Factors affecting growth rate

In the final column in Table 3(a) the normal duration of larval development for the three insects concerned is given and it will be seen that the values vary in the same order as the relative growth ratios. Obviously to some extent the time for development depends on the amount of growth to be made but even if allowance is made for this the growth rates still clearly differ. One factor that almost certainly is additionally responsible is the quality of food available to the insects concerned. Blow fly larvae normally live in a nutritious food medium (some infest wounds on living animals). Honey bee larvae are within a nest and are fed regularly by worker bees.

* — *Growth* of an organism is increase in size of the body as a whole.
— *Development* implies differential growth so that some parts of the body grow more than others resulting in a change of form.
— *Metamorphosis* involves a marked change in bodily form between one stage in the life cycle and another.

Goat moth larvae on the other hand are wood boring and consequently are on a rather poor nutritional plane. Food quality (and quantity) therefore can have a marked effect on insect growth rates and this applies within species as well as between different species.

Table 3. (a) Growth ratios and comparative growth rates of three insects

Larva of	*Weight increase —egg to mature larva*	*Normal development time (natural conditions)*
blow fly	× 450	70 hours
honey bee	×1500	5 days
goat moth	× 72,000	3 years

(b) Effect of temperature on rate of growth

Temperature °C	*Days from egg hatch to pupation* house fly	white butterfly
10	34	55
20	9	15
30	5	8

Data of Imms, 1947

A further factor affecting insect development, and at least as important as food, is temperature. The effect of temperature on the rate of development of two common insects is shown in Table 3(b). Insects are unable to regulate their body temperature to any extent and for the most part take on the temperature of their surroundings. As all bodily activity is temperature dependent, including biochemical activity on which growth depends, the lower the temperature the slower an insect's rate of growth. Increased temperatures of course have the opposite effect. In the examples given in part (a) of Table 3, environmental temperatures may also have contributed to the different developmental times. Blow fly larvae which inhabit wounds on warm blooded animals would benefit from their body temperature, and honey bees by communal activity do to some extent regulate temperature within the hive. However, goat moth larvae, boring within the wood of a tree have no such advantages.

Although both insects in Table 3(b) showed maximum growth rate at 30°C it must not be assumed that temperature can be increased indefinitely without harmful effects. The optimum for most insects is in the range 30-35°C, with an extreme upper limit of about 40°C. Similarly, at the lower end of the scale, most insects become inactive, do not feed and cease to grow below 10°C. Temperatures below 0 °C are quickly lethal to most insects and many of those that survive cold winter conditions with frost and snow do so because they are well protected in some way rather than because they can withstand freezing. Some insect species however do have the ability to tolerate temperatures below freezing point.

Moulting (Ecdysis)

Some advantages conferred by an external skeleton were given in Chapter 1. There is however one serious disadvantage. The harder parts of the exoskeleton have negligible ability to stretch so that as an insect grows it must shed its skin at intervals and grow a new one in a larger size, a process known as **moulting** or **ecdysis**. This feature is common to all animals with an external skeleton, ie, all arthropods, not only insects. We find therefore that the hard parts of an insect increase suddenly in size after each moult so that their growth proceeds in a series of jerks. Between moults body weight increases and the softer parts of the exoskeleton stretch, but when this reaches its limits moulting must again take place. Each growth stage of an insect between moults is referred to as an **instar**. The process is shown diagrammatically in Fig 11.

Some insects undergo very few moults, though never less than three. House fly larvae for example regularly pass through three instars. More commonly however there are rather more than this: 7-9 in cockroaches for instance. The highest number (30-40) is found in mayflies (Ephemeroptera). In general the more primitive insect groups undergo a large and variable number of moults whereas those that are more highly evolved have fewer moults and fixed in number for the species.

It has been known for many years that the initiation of moulting in insects is due to a chemical factor (a hormone) released into the insect's blood (haemolymph). It was early given the name **moulting hormone** (often now abbreviated to **MH**) and is sometimes also known as **ecdysone**. Its chemical identity has been established in recent years and found to be a steroid, as are many mammalian hormones. The same chemical appears to be utilised by all insects and perhaps even by all arthropods. Its site of production has been identified as a gland present in the fore part of the thorax of insects, the **prothoracic gland.** Between moults the concentration of moulting hormone in the insect body is low but just prior to moulting it is released in large amounts.

Some synthetic chemicals are now known which interfere with moulting in such a way that the insect dies before the process is successfully completed. The utilisation of such substances as insecticides is discussed in Chapter 11.

Metamorphosis

Perhaps the most striking feature of insect development is the marked change of bodily form which most undergo from young to

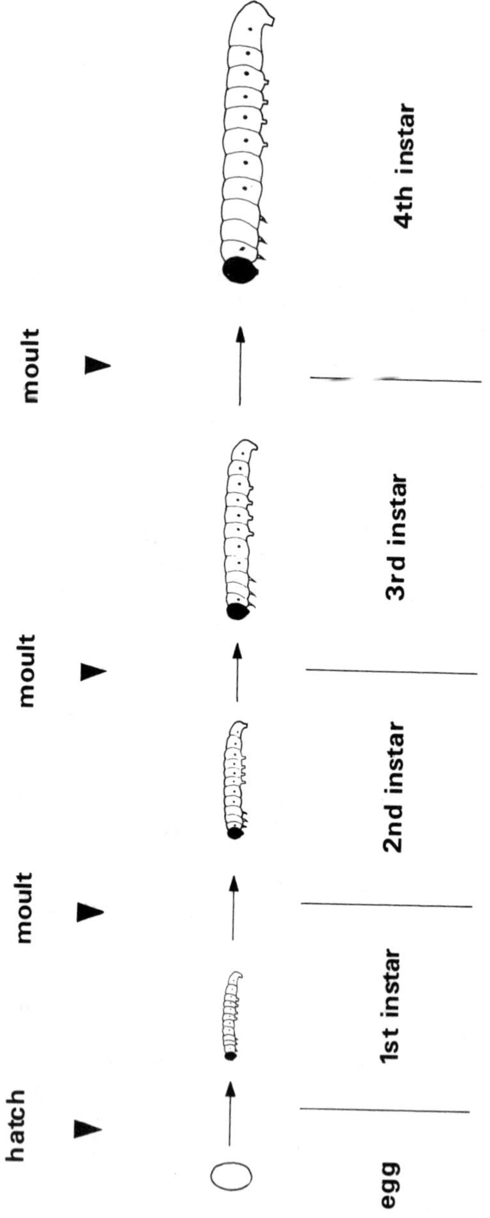

Fig. 11. The pattern of growth of a typical insect.

adult, a process known as **metamorphosis**. Everyone is probably familiar with the dramatic transformation which a caterpillar undergoes as it changes to a butterfly or moth via a **chrysalis (pupa)** as depicted in Fig 12(b). Many other insect groups (beetles, flies, wasps, etc) undergo similar transformations. In such cases, where the immature stages are totally different in appearance and life style from the adults, the process is referred to as **complete metamorphosis**. The correct term for the juvenile stage in such cases is **larva** and that for the adult, **imago**. The stage between larva and imago, when transformation is taking place, but which is not active and does not feed, is the **pupa**. The term **chrysalis** is sometimes applied to the pupa stage (particularly that of butterflies). A **cocoon** is the silken covering which some insects form to protect the pupa.

Complete metamorphosis evidently imparts some biological advantages as it occurs in all the more highly evolved insect groups. It enables the insect to exploit two different environments according to life stage. Larvae for the most part are just eating machines and are not very mobile. They are often highly adapted to their food and to their niche in the environment, and this is when rapid growth takes place. The role of the adults is one of reproduction and dispersal. Adult insects are usually winged and thus capable of moving considerable distances. Once an insect reaches adulthood, growth and moulting cease (except for very primitive insect groups).

Not all insects undergo the drastic type of transformation described above. In groups such as crickets and cockroaches the juvenile stages are rather like small versions of the adult and transformation to the adult form takes place gradually with each succeeding moult (Fig 12(a)). There is no pupal stage and with the last moult the insect becomes a fully fledged adult. However the juvenile stages are not *identical* to the adults, even allowing for size. In juvenile insects for example the wings and external reproductive organs are not fully formed, but gradually assume the mature form as the insect proceeds from one instar to the next. There is thus some degree of metamorphosis and this type of development is referred to as **partial** or **incomplete metamorphosis**. It is the more primitive condition from which complete metamorphosis has evolved. The juvenile stages of insects showing this type of development should strictly be referred to as **nymphs**, but unfortunately the term larva is now often used indiscriminately for any immature insect.

Wing development takes place in a different manner in the two types of metamorphosis. In insects with complete metamorphosis, there are no external signs of wings at all in the larval stages. Wings develop internally within the pupa (in a folded condition) so that

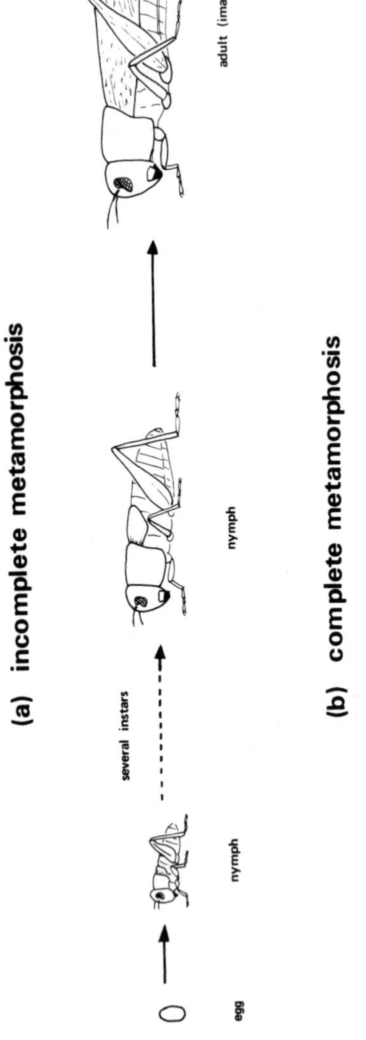

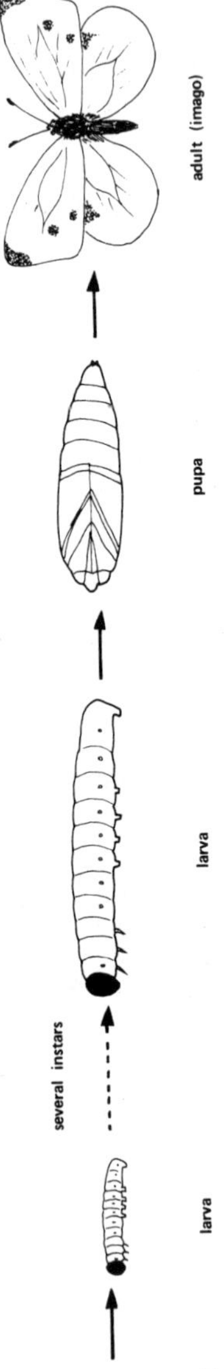

Fig. 12. Types of insect metamorphosis.

when the adult emerges it is fully winged. In contrast, with incomplete metamorphosis the wings develop externally and first appear as small outgrowths or wing stubs in an early instar. These gradually enlarge with each succeeding moult until the wings are fully formed after the final moult.

Besides the intrinsic interest of these two types of development there is an important practical consideration as far as plant pests are concerned. This concerns the fact that the feeding habits of larvae and adults of most insects undergoing complete metamorphosis are totally different. The larval stage is the main, and in most cases the only stage causing plant injury. The main exceptions are some of the beetles (Coleoptera) where both adults and larvae may be plant damaging. In insects with partial metamorphosis however, the adults and immature stages possess the same mouthpart pattern and feed in an identical fashion so that both stages are plant damaging.

Some primitive insects, such as springtails (Collembola), show no metamorphosis at all but simply increase in size from juvenile to adult. Also they may continue to moult once adult. Furthermore, none of these insects possesses wings and they are believed to be descended from ancient insect stock before wings had evolved.

It is extremely important to have an understanding of these different types of development as they are used as the basis for classifying the insects into major subdivisions, as indicated in Table 4. Those primitive insect groups which show no metamorphosis are placed in the sub-class **Ametabola**. As they also are wingless an alternative name (in brackets in the table) is **Apterygota**. All remaining insects, that is those showing some metamorphosis whether complete or incomplete, are placed in the **Metabola** (or **Pterygota**). The Metabola are further subdivided into two divisions, the **Hemimetabola** for those groups showing partial metamorphosis, and the **Holometabola** for those with complete metamorphosis. Alternative names are **Exopterygota** for the Hemimetabola (as wings develop externally) and **Endopterygota** for Holometabola (as wings develop internally in the pupal stage). Included in the right hand column of Table 4 are the main characteristics of each group. Further reference is made to these major subdivisions in Chapter 6 where insect classification is discussed in more detail.

Just as moulting is regulated in insects by a hormone so also is metamorphosis. In this case the hormone responsible is **juvenile hormone (JH)**. In contrast to moulting hormone it acts in a negative fashion so that its continued presence suppresses metamorphosis while its absence is what allows metamorphosis to proceed. Juvenile hormone is produced by special glands, the **corpora allata**, which are closely associated with the brain. As with moulting hormone, the chemical nature of juvenile hormone has been elucidated in recent years, and found to be a terpenoid, in contrast to the steroidal nature of moulting hormone.

Both juvenile hormone itself, and many chemically related substances with strong juvenile hormone activity have been synthesised in the laboratory. Some of these synthetic materials actually have greater activity than the natural hormone and can be taken up by the insect following external application. If an insect is treated with a juvenile hormone substance in this way at an inappropriate time in relation to its development, its normal progress can be severely disrupted so that later it dies. In effect these materials can be used as insecticides and as such are discussed further in Chapter 11 where chemical control of insect pests is considered.

Table 4. The major divisions of the insects and their essential features

Sub-Class	*Division*	*Main features*
Ametabola (Apterygota)		No appreciable change in form with growth (though smaller stages not sexually mature). Moulting may continue once adult. Wings never present.
Metabola (Pterygota)	Hemimetabola (Exopterygota)	Metamorphosis incomplete. Pupal stage absent. Wings develop externally. Young stages (nymphs) similar in form to adults; possess similar mouthparts and feed in the same fashion. Nymphs may possess compound eyes.
	Holometabola (Endopterygota)	Metamorphosis complete. Pupal stage present. Wings develop internally (in pupa). Young stages (larvae) very different in form from adults; possess different mouthparts and feed in a different fashion. Larvae never possess compound eyes.

SELECTED REFERENCES

Chapman, R. F. 1982. *The insects: structure and function*. 3rd edition. Hodder and Stoughton/Harvard University Press. 832 pp.

Imms, A. D. 1947. *Insect natural history*. Collins, London. 317 pp.

Richards, O. W.; Davies, R. G. 1977. *Imm's general textbook of entomology*. 10th edition. Vol. I. Structure, physiology and development. Chapman and Hall, London. 418 pp.

Ross, H. H; Ross, C. A. and Ross. J. R. P. 1982. *A textbook of entomology*. 4th edition. John Wiley and Sons, New York. 666 pp.

Wiggelsworth, V. B. 1972. *The principles of insect physiology*. 7th edition. Chapman and Hall, London. 827 pp.

Chapter 5

Reproduction and Life Cycles

A prominent feature of the biology of most insects is their ability to multiply rapidly under favourable conditions. Two components contribute towards this, namely the large numbers of eggs produced per female and the short life cycle exhibited by many insects. Although the potential reproductive capacity of most insects is prodigious it is counterbalanced in nature most of the time by high mortality. The natural factors which regulate insect populations are discussed more fully in Chapters 9 and 10. The actual reproductive process is considered here.

For most insects reproduction depends on mating between the sexes, followed by egg laying. There are however a number of important departures from this basic pattern. The normal sexual mode of reproduction is considered first in some detail then the main variations exhibited by some insects are discussed.

The normal sexual reproductive cycle

The reproductive cycle may be considered as a series of sequential steps:

1. Sexually mature adults
2. Locating the opposite sex
3. Mating
4. Pre-oviposition period
5. Egg laying (oviposition)
6. Egg hatch
7. Development of immature stages to adulthood.

Once immature stages have completed their growth and development and are sexually mature the cycle is of course complete. Each step in the cycle may now be considered in detail:

1. *Sexually mature adults*

After an insect has undergone its last moult and emerged as a fully fledged adult there is usually a period of maturation before it is capable of mating and laying eggs. This period may be quite brief for those insects with short lived adults, especially in species that do not feed at this stage. However, with many insects the delay is longer and it may be essential for the insect to feed before it is sexually mature. Female mosquitoes for example normally must take a blood meal before eggs can mature properly. Therefore for many insects quantity and quality of food available to the adult influences the number and viability of eggs produced.

2. *Locating the opposite sex*

As most adult insects are highly mobile and may disperse widely, special sex attractant mechanisms have evolved to enable the sexes to locate each other. Some insects, such as crickets and cicadas, use sound for this purpose and special sound producing systems are present. These may involve rubbing one part of the body against the other, eg, one forewing over the other as in some crickets, or a leg against a wing as in some grasshoppers. In other cases sound provided by the wing beat may be utilised, as in mosquitoes. The most sophisticated sound producing organs occur in cicadas which possess a special membrane stretched across a sound chamber on either side of the lower surface of the abdomen. The pattern of sound produced by flexing of the membrane is distinctive for each species. Sound production in cicadas is confined to the male and it is the female that responds. In other insects it may be the other way round.

In insects that use sound as a sex attractant mechanism, sound receiving organs must be present, at least in the responding sex. These usually consist of distinct structures incorporating a membrane of some sort which is set in vibration by the sound impulses. In crickets the sound receiving organ can be seen as a light oval patch on the centre part of the foreleg.

The use of sound for attracting the opposite sex however is not common and most insects use chemical scents instead. These are the **sex attractant chemicals** (**sex pheromones**). They were first discovered in moths many years ago, but are now known to occur in most insect groups. They are nearly always produced by the female to attract the male but it can be the other way around, as in the cotton boll weevil. Sex pheromones of insects are extremely potent and may be able to attract the opposite sex over long distances. With some moths for instance the range may be several kilometres but for most insects it is considerably less than this, perhaps a few metres. The need for long distance operation depends of course on the biology and behaviour of the species concerned.

Besides being potent, insect sex attractants are extremely specific, usually attracting only males of the same species or at most

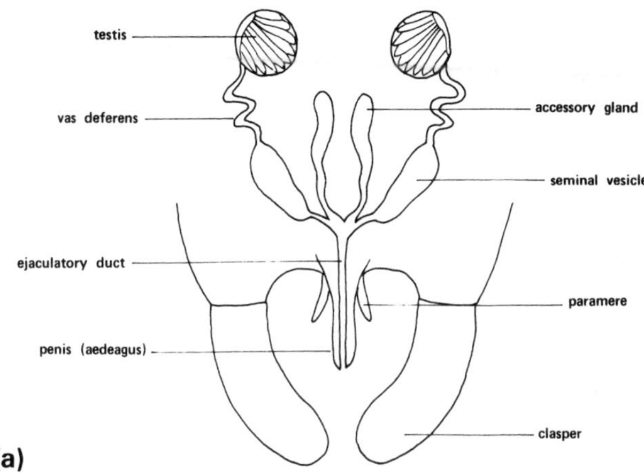

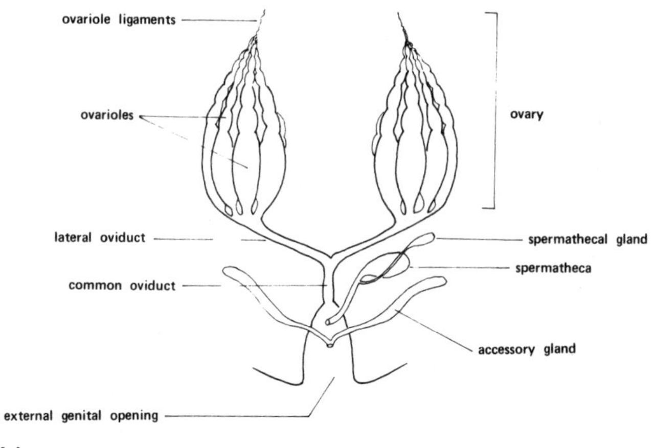

Fig. 13. (a) The structure of male insect reproductive organs (diagrammatic).
(b) The structure of female insect reproductive organs (diagrammatic). (both after Snodgrass).

a very few closely related species. Many have now been chemically identified and some have been synthesised. Often they consist of mixtures of compounds rather than single substances. Chemical insect sex attractants have valuable practical applications in surveys and monitoring, and possibly for control of some species. These aspects are discussed further in Chapter 11 where chemical control of insect pests is considered.

Some insects show marked differences in appearance (size, shape, colour) between the sexes, a phenomenon known as **sexual dimorphism**. This is particularly prominent with some butterflies and in such cases a visual component may be involved in sex attraction.

3. *Mating*

In order to understand the reproductive process of insects, particularly that part concerned with mating, the structure of the reproductive organs must be considered. This was touched on briefly in Chapter 3 which dealt broadly with insect anatomy. The general arrangement of the reproductive organs of both sexes is shown diagrammatically in Fig 13.

In the male the internal organs consist of a pair of **testes** in which **sperm** is produced, though in some insect groups, eg, some Lepidoptera (butterflies and moths), they are fused into a single structure. From each testis a duct (**vas deferens**) leads down to a storage reservoir for sperm (**seminal vesicle**). From each seminal vesicle a further duct connects to the male copulatory organ (**penis, aedeagus**) via a central **ejaculatory duct**. Paired **accessory glands** usually open into these latter ducts below the seminal vesicles. The exact function of the accessory glands in male insects is uncertain but there is evidence that their secretions in some cases inhibit further mating of inseminated females, and also stimulate oviposition.

Externally at the tip of the male abdomen there is usually a pair of **claspers** of varying shape which grasp the abdomen of the female during copulation.

In female insects the internal reproductive organs are paired **ovaries** that lie within the body cavity of the abdomen. Each ovary is made up of a number of similar units called **ovarioles** (see Fig 13 (b)). The number of **ovarioles** varies greatly in different insects depending on egg laying capacity. Tsetse flies, which produce few eggs, have only one ovariole per ovary, whereas the queens of social insects which produce very large numbers of eggs may have more than a hundred ovarioles in each ovary. Each ovariole is like a tapered tube. New eggs are budded off at the top and as they pass down they are nourished and grow so that by the time they reach the base they are mature and ready to be fertilised and laid.

From each ovary a lateral oviduct leads to a single central **common oviduct**. Opening into this there is a distended sac-like side branch, the **spermatheca**, which usually has a special gland (**spermathecal gland**) associated with it. In addition, paired accessory glands open into the central common oviduct. In most insects the external opening of the oviduct, besides providing for deposition of the eggs, also serves for reception of sperm from the male during copulation, but some insects have a separate opening for this purpose as in the Lepidoptera (butterflies and moths). Externally, most female insects possess an ovipositor to assist with egg laying, as discussed in Chapter 3.

Egg development in unmated female insects proceeds to the stage where mature eggs are present at the base of each ovariole. Here they are normally held until after mating has taken place. During mating, sperm from the male is received into the spermatheca from which it can be released to fertilise eggs when they pass down the common oviduct. The spermathecal glands are probably concerned with retention of sperm in a viable condition within the spermatheca. This may be for considerable lengths of time in long lived insects such as queen bees which mate only once during their life time. The accessory glands of the female reproductive system provide a secretion which coats the eggs as they are laid. In some insects this may also involve sticking the eggs to a surface of some sort; for example, many Lepidoptera (butterflies and moths) cement their eggs to the leaves of plants suitable for their larvae to feed on. In a few cases clusters of eggs are glued together to form a distinct structure, the **ootheca** or **egg purse**, as produced by mantids and cockroaches.

Mating in some insects, eg, many moths, involves the formation of a package (**spermatophore**) enclosing the sperm. This is usually received into a special sac within the female reproductive tract (**bursa copulatrix**) from which sperm is later transferred to the spermatheca.

4. *Pre-oviposition period*

Egg laying (**oviposition**) does not usually commence until some time after copulation. This period of delay is referred to as the **pre-oviposition period**. In short lived insects it may be very brief so that egg laying commences almost immediately after mating has taken place. Some insects need to mate more than once if the full reproductive potential is to be achieved but in others a single mating may suffice even though the female may be quite long lived.

5. *Egg laying (oviposition)*

Normal egg production for most species of insects is at least a hundred per female, but is commonly much more. Eggs may be laid singly or arranged in groups, sometimes with a definite pattern characteristic of the species.

Plant feeding insects usually lay their eggs on plants which provide suitable food for the young stages when they hatch, but this is not always the case. Some species simply scatter their eggs in areas where suitable food plants are likely to be available. In such cases the immature eggs are likely to be capable of feeding on a wide range of plants otherwise their chance of survival would be poor.

The form of insect eggs varies considerably. Some are almost spherical but others are flattened (eg leafrollers), or elongate (most flies (Diptera)). There is sometimes a definite shell to the eggs as in butterflies and moths but in other cases, as for example with some soil inhabiting beetles, the eggs are quite soft and must absorb moisture before hatching. Except for social species (those living in large and complex colonies) insect eggs are abandoned once they are laid and the young on hatching have to fend for themselves.

6. *Egg hatch*

When most insect eggs are laid embryological development has hardly started so that a period of time must elapse after laying before the eggs hatch. The duration of this period is very dependent on temperature. Success of development and egg hatch may also depend on adequate humidity, and absorption of water may be essential for development of the eggs of some insects. The egg provides a very convenient stage for insects to pass the winter in colder climates (see end of this Chapter) and many pests of annual and deciduous plants adopt this strategy.

7. *Development of immature stages to adulthood*

The general patterns of insect growth and development were considered in Chapter 4 which dealt in detail with the development of immature stages to adulthood and need not be further discussed here. One additional point however is that the reproductive capacity of adult insects may depend greatly on the quantity and quality of food available to the *immature stages*, particularly for those species that do not feed once they reach adulthood.

Life cycle diagrams

Insect life cycles can be conveniently depicted in the form of a diagram of the type illustrated in Fig 14, which shows the normal pattern of sexual reproduction for (a) insects with incomplete metamorphosis and (b) insects with complete metamorphosis.

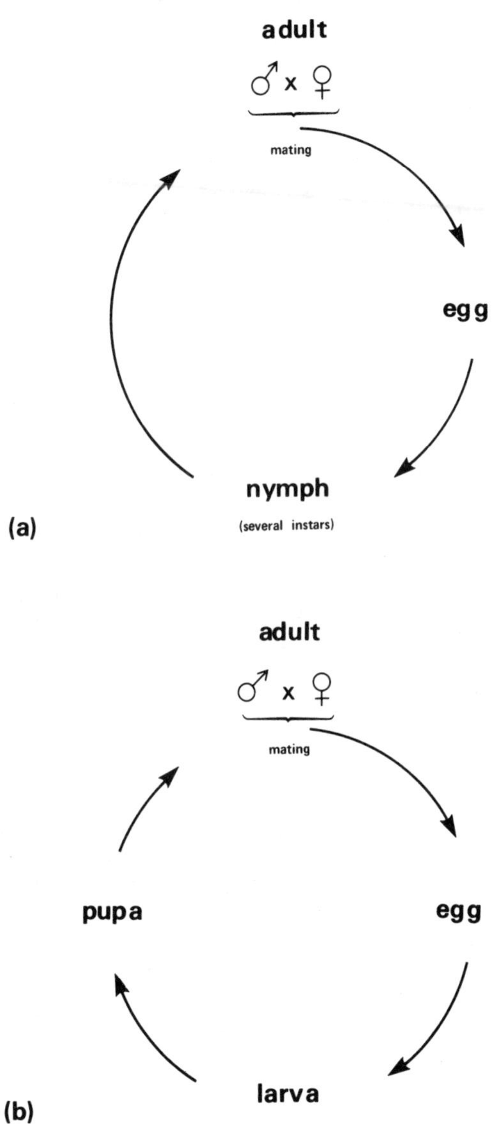

Fig. 14. Life cycle diagrams of insects with sexual reproduction and (a) incomplete metamorphosis, eg, crickets, plant bugs, (b) complete metamorphosis, eg, beetles, moths.

Variations to the normal pattern of reproduction

Departures from the normal pattern of reproduction are quite common. The more important of these are:

(a) *Parthenogenesis*

This involves the ability of female insects to produce viable eggs without fertilisation by the male. Such eggs develop normally and the remainder of the life cycle is unaffected. There are basically two types of parthenogenesis in insects.

In some species it is the only mode of reproduction, and in such cases males are entirely unknown. An example is provided by white fringed weevil where eggs (laid in the usual way) all give rise to females. All individuals are diploid (having chromosomes present in homologous pairs). The process is depicted in Fig 15(a).

In other cases parthenogenesis is optional, with fertilised eggs giving rise to females which are diploid while unfertilised eggs still develop but give rise to males which are haploid (having a single set of unpaired chromosomes). The process is shown diagrammatically in Fig 15(b). This mode of reproduction is common amongst parasitic Hymenoptera (wasps) and has the advantage that females which fail to find a mate can still reproduce. In the social Hymenoptera (ants, and most bees and wasps) the reproductive females (queens) are mated but are able to regulate fertilisation of eggs as they are laid so that male and female offspring can be produced as required.

(b) *Viviparity*

In this condition fertilised eggs are retained in the oviduct of the female until they hatch and active young larvae or nymphs are deposited rather than eggs. This occurs in some species of blowflies. In most such cases there is no nourishment of the young, the eggs simply being held within the common oviduct until hatching occurs. However, in a few insects nourishment is provided for the young insects after they hatch. The extreme of this is shown by the tsetse fly, the larvae of which are nourished within the mother so that when they are deposited they do not need to feed and promptly pupate.

(c) *Parthenogenesis and viviparity*

In a few insects parthenogenesis and viviparity occur together. The best known example is provided by aphids during summer. All individuals in the population are female and they give rise to living young without mating. As aphids also undergo incomplete metamorphosis the young are similar in appearance to the adults and the

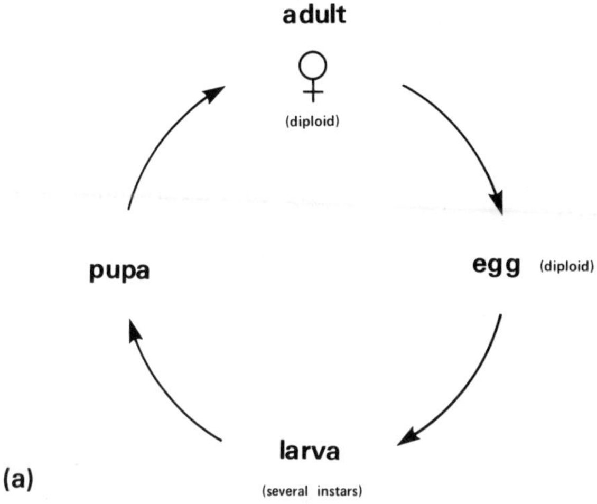

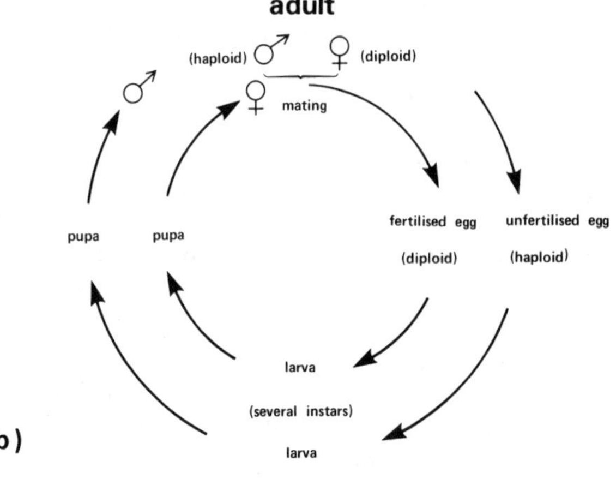

Fig. 15. Life cycle diagrams of insects with parthenogenetic reproduction;
(a) all offspring diploid and female, eg, white fringed weevil,
(b) unfertilised eggs give rise to male offspring (haploid), fertilised eggs give rise to female offspring (diploid), eg, many Hymenoptera (ants, bees, wasps).

life cycle is thus very simple. (*Note*: In cold climates aphids normally produce males and females at the onset of winter and eggs are then laid which overwinter.)

(d) *Polyembryony*

This is a special reproductive device that only occurs in a few insect species and functions as an insurance for survival. A single egg, instead of giving rise to one embryo, divides and subdivides several times before embryonic development commences so that a number of young are produced. Polyembryony occurs particularly in some parasitic wasps where several hundred young may develop from one egg deposited into the parasitised host insect.

Generations a year

There is a great deal of variation between insect species in the length of the life cycle and hence the number of generations a year. With some insects, such as codling moth in cool climates, the life cycle takes twelve months so that adult moths appear just once during the year. Far more commonly however (as with white butterfly) there are several generations per year and for most of the time the different life stages overlap. A few insects have very extended life cycles and take several years to develop. This is the case with cicadas most of which spend 3-5 years as nymphs but as there are different age groups some individuals emerge each season. The longest lived insect of all is in fact a cicada, the 17 year (or periodical) cicada of North America which takes up to 17 years to develop.

At the opposite extreme, the life cycle of aphids may take less than a week under summer conditions so that build up of numbers can be very rapid. Many important crop pests have short life cycles and the associated ability to multiply rapidly.

Overwintering

Moist tropical climates are favourable for insect development all the year round and reproduction is therefore usually continuous with generation succeeding generation indefinitely. Besides providing suitable environmental conditions, the tropics also ensure an adequate food supply at all times of the year, except where severe dry seasons occur.

In contrast, temperate climates are unfavourable for insects during the winter as temperatures are too low for most insect activity and furthermore, the food supply of insects that depend on annual or deciduous plants is not available. Most insects in temperate climates therefore pass the winter (overwinter) in a dormant or semi-dormant state. This can be any stage in the life cycle but is commonly the egg or pupal stage as these do not require food and can remain in an inactive condition for long periods. During the winter many insects enter a specialised condition of suspended activity called **diapause**. This is often initiated before the onset of severe adverse conditions by some environmental trigger (such as decreasing daylength) so that by the time winter strikes the insect is already dormant. Diapause also normally requires a particular stimulus for its termination, such as a certain period of time at cold temperatures. This is clearly a device to prevent the insect being aroused by a brief spell of warm weather and then being caught out when the temperature drops again. Diapause can occur at any stage in the life cycle but is most common in the larval or pupal stage.

The practical relevance of knowledge of life cycles

Knowledge of the life cycle of a pest enables one to anticipate when and where particular life stages will occur. If we understand the sequence of stages and their duration then the time when plant damage may eventuate can also be determined. The ability to recognise early stages in the life cycle is important in the latter respect. For instance the appearance of eggs of white butterfly (which are easily recognised) on the leaves of cruciferous plants indicates that damage from feeding larvae will follow shortly.

Perhaps more importantly, knowledge of the life cycle enables control measures to be timed more effectively. For instance grass grub (a serious pest of pasture in New Zealand) can be severely affected by cultivation of the soil when in the pupal stage. For maximum effect most of the population must be present as pupae, which means early spring for most districts. Accurate timing of insecticide applications is also important for many pests. Codling moth for example lays its eggs mostly on the leaves of apple trees adjacent to developing fruits. The young larvae on hatching must then move to a fruit and tunnel in to ensure survival. Control depends on application of an insecticide to kill the newly hatched caterpillars before they become established inside fruits. Timing is important as too early an application will be wasteful whereas treatment made too late will be ineffective against those larvae that have already entered fruits. Unfortunately, under British conditions egg hatch of codling moth is rather spread in time making the problem a difficult one.

Finally, knowledge of the life cycle may enable vulnerable points to be identified and may facilitate prediction of outbreaks. For example, the black bean aphid in Britain passes the winter as eggs which are laid on spindle trees in hedgerows and surrounding areas. Counts of such eggs during the winter provide estimates of likely spring populations. In New Zealand on the other hand aphids rarely produce sexual forms or lay eggs so that this approach is not feasible.

These examples do not by any means exhaust the range of situations where knowledge of insect life cycles has important practical implications. In general we may state that adequate knowledge of pests' life cycles is one important component of information about their biology in general that is crucial to their rational control.

SELECTED REFERENCES

Chapman, R. F. 1982. *The insects: structure and function*. 3rd edition. Hodder and Stoughton/Harvard University Press. 832 pp.

Jacobson, M. 1972. *Insect sex pheromones*. Academic Press, New York. 382 pp.

Richards, O. W.; Davies, R. G. 1977. *Imm's general textbook of entomology*. 10th edition. Vol. I. Structure, physiology and development. Chapman and Hall, London. 418 pp.

Ross, H. H.; Ross, C. A.; and Ross, J. R. P. 1982. *A textbook of entomology*. 4th edition. John Wiley and Sons, New York. 666 pp.

Snodgrass, R. E. 1935. *Principles of insect morphology*. McGraw Hill, New York, 667 pp.

Chapter 6

Insect Identification and Classification*

Insect identification and classification are two closely related but not identical topics. Identification of insects (and other organisms) is concerned with putting names to individual species. Classification on the other hand involves the arrangement of such species in some orderly fashion. To many people, identification of insects is a tedious and academic exercise which has little practical value. This is in fact far from the truth. Accurate identification of a pest is the first and perhaps the most important step to be taken in dealing with it as the name provides the key to all published information. Without a name this information is not accessible. Furthermore, identification must be accurate to be of value. If an insect is misidentified, ie, we think it is one species when in fact it is actually something else, any information obtained will be wrong as it will apply to the wrong organism. It is preferable therefore not to make an identification at all than to make an incorrect one.

Insect classification also has some practical spin off. Species of insects are so numerous that some orderly arrangement is necessary to bring order out of chaos and to allow us to find our way around in a morass of detail. Groups within a good system of classification have certain elements in common so that if we know something about one member we can expect others to be similar. No system of classification of living organisms is perfect however and revision of detail is being continually made with advancing knowledge. Only the broad outlines of insect classification are included here.

Insect names

The system of scientific names applied to insects is that which is now used universally for almost all groups of living organisms. This is the **binomial system** of latin names devised by the Swedish naturalist Linnaeus some 200 years ago. Under this system each

* The study of the naming and classification of organisms is referred to as **taxonomy** or **systematics**.

species is given a double name. The first name, which is always written with a *capital* initial letter is the **genus** (generic name). This may be likened to a human surname. The second name, which is always written with a *small* initial letter is the **species** (specific name), akin to a human Christian name. Each genus usually includes several species (as there are usually several members to human families), but there may be only one. The great advantage of the binomial system of latin names is that it is internationally recognised so that the same insect, which may occur in many parts of the world, is known by the same name. This would certainly not be the case if common or colloquial names were used as the same insect may be known by several different names in different countries. There is also the problem with common names that two (or more) similar insects may be referred to by the same name. If common names for insects can be standardised however, they are very useful as most people are familiar with them. In Britain the Ministry of Agriculture, Fisheries and Food has published a comprehensive list of common and scientific names for invertebrates of economic importance (see references at the end of the Chapter), and similar lists are available in several other countries.

As an example of an insect scientific name we may take that of cabbage root fly which is —

genus species	*author*
Delia brassicae	(Wiedemann)

capital initial letter — small initial letter

Scientific names in publications are usually printed in italics, or underlined in the case of typescript.

In the case of cabbage root fly the genus is *Delia* (with a capital initial letter) and the species is *brassicae* (with a small initial letter). The final name of Wiedemann is that of the person who first described and named this particular insect. In this instance, however, when Wiedemann recognised the species *brassicae* he placed it in a genus other than *Delia**. Later revision suggested that it did not properly belong there and thus it was transferred to its present genus, *Delia*. This explains why the name Wiedemann is enclosed in brackets in the above example. Full citation of scientific names should include the author but this is often ommitted for brevity.

Means of identifying insects

There are several ways of dealing with the problem of identifying insects.

* This insect has in fact undergone several name changes.

(a) *Use of keys*

Keys are published aids to the identification of insects (and other organisms) which use a sequence of steps based on diagnostic characters of the species concerned. Many readers will be familiar with such keys for the identification of plants. In theory the use of keys to identify insects is attractive, but in practice there are two major problems. Firstly, the availability of keys to various insect groups is often limited. Fortunately in Great Britain detailed keys have been published for all the important insect groups (see references at the end of this Chapter), but in many countries this is not the case and often only a few keys to various segments of the insect fauna are available. This situation will of course improve with time.

Secondly, many of the details of insect anatomy and associated terminology, which keys to species level must perforce employ, are not familiar to the average person. A further complication is that often the features used are small so that access to a stereo-microscope is essential.

(b) *Identification services*

When correct identification of an insect is important assistance should be sought from the Agricultural Development and Advisory Service of the Ministry of Agriculture, Fisheries and Food. Their regional centres retain specialist staff who will identify and provide advice on pests, plant diseases and also weed problems. Staff of museums in some centres may also undertake insect identification. It is important that such services should not be overloaded with trivial requests for identifications which, in the main, should be limited to those of economic importance. Authoritative identification is particularly important if it is suspected that a pest species new to the country has been found.

(c) *Visual recognition*

All agriculturalists and horticulturalists should get to know the important pests that affect the plants with which they are concerned so that they can recognise them on sight. This is not too difficult as with familiarity insects become as distinct as human beings. It is true that some closely related species are difficult to distinguish even by the specialist, but this is the exception rather than the rule. Plant damage symptoms are a valuable aid to identification as they are often highly characteristic of the pest concerned. For instance, in New Zealand an apple that has the core eaten out must be due to codling moth as no other pest occurs there which causes such injury. In Britain, however, similar injury may be caused by apple sawfly.

One cannot claim complete familiarity with an insect until all stages in the life cycle can be recognised. This may be difficult for eggs or pupae, but sometimes there are characteristic features that enable their identity to be established. The location of such stages, for example in the soil or on particular plants, may also provide valuable clues. Many pests cause plant injury only in the larval stage but ability to recognise the adults as well may be most important in practice, as this may determine timing and placement of control measures. Timing of sprays against codling moth for instance should preferably be based on peaks of appearance of the adult moths.

With general visual recognition of pests, confusion is most likely to arise where there are two or more species of similar overall appearance and differing only in detail. Such is the case with several species of aphids that attack potatoes for example. Some differences between such species do invariably occur however and attention to detail will enable them to be separated.

Higher and lower categories (than species)

Species of organisms are to a great extent the only units of classification that really exist. All other categories, particularly the higher ones, are largely abstractions invented for our own convenience. Classification systems for organisms are hierarchies; that is, each category at a particular level encompasses a number of units from the level below. Similarly, these again embrace a number of units from the next lower level and so on.

Although classification systems of organisms, including insects, are largely theoretical, a good classification is based on degrees of relatedness and should thus have some evolutionary significance. With insects, however, the fossil record is poor and we cannot be certain of the origin of many of the groups of modern insects.

Three main categories are recognised above the species level. Thus species are grouped into **genera**, genera into **families**, and families into **orders**. The classification of the garden chafer (*Phyllopertha horticola*) in this respect is given in Table 5.

The orders of insects comprise the main insect types, eg, beetles—order Coleoptera, flies—order Diptera, butterflies and moths—order Lepidoptera. In the larger orders several other categories are inserted at an intermediate level to provide for breakdown into manageable portions, eg, sub-order, super-family, tribe.

About 30 insect orders are recognised but the exact number will depend on which authority is followed. In this book the classification used is that of Imms (10th edition, 1977). An outline of the classification of insects is set out in Table 6. The grouping of orders into sub-classes is based on the type of metamorphosis (if any), as discussed in Chapter 4. For most insect orders there is a common name equivalent, as in the examples given above, but this is not always the case as for example with the Orthoptera. Further consideration of the orders of insects is included in the final section of this Chapter.

Although 29 insect orders are listed in Table 6, only about half that number are of economic importance (as indicated in the Table) and if we restrict interest to those associated with plants the number is reduced even further.

Sometimes categories below the level of species are recognised, such as subspecies, variety, strain, race. These may be based on minor differences in bodily structure or colour or may have a physiological basis, such as ability to utilise different host plants. Such differences may be quite important with some economic insects. All such groups interbreed freely given the opportunity, whereas different species normally do not.

Table 5. An example of the classification of an insect—the garden chafer *Phyllopertha horticola*

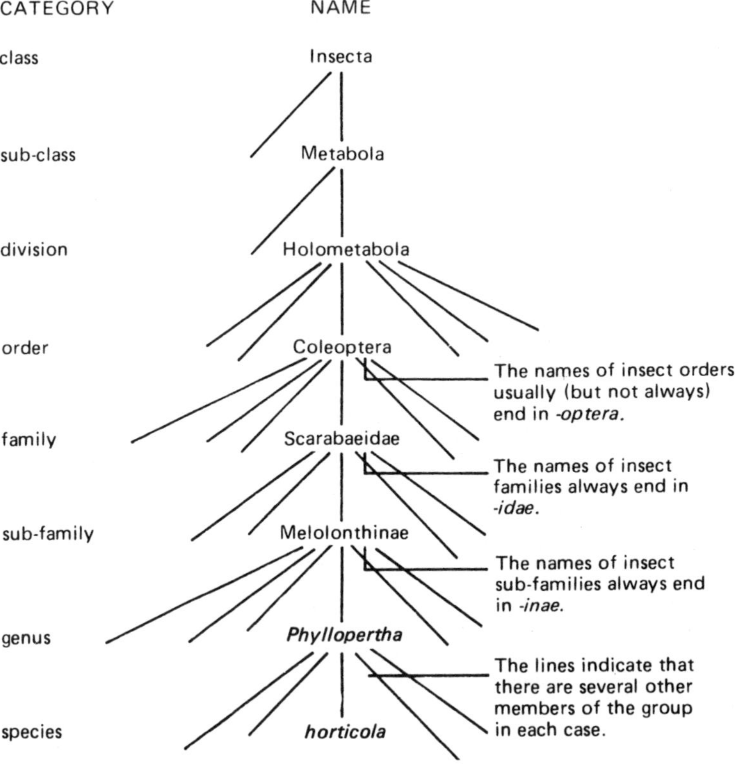

Table 6. An outline classification of insects

subclass I AMETABOLA (= APTERYGOTA).

Adults wingless (primitive condition). No metamorphosis. Moulting continues once adult.

	Order		Common names
order 1.	Thysanura		— three-pronged bristle tails, silverfish
2.	Diplura		— two-pronged bristle tails
3.	Protura		— proturans
4.	Collembola	*	— springtails

subclass II METABOLA (= PTERYGOTA)

Adults usually winged (but may be secondarily wingless). Metamorphosis occurs. Do not moult once adult (except for mayflies).

division A. HEMIMETABOLA (= EXOPTERYGOTA)

Metamorphosis incomplete. No true pupal stage. Wings develop externally in nymphal stages.

	Order		Common names
order	5. Ephemeroptera		— mayflies
	6. Odonata		— dragonflies
	7. Plecoptera		— stoneflies
	8. Grylloblattodea	ø	— —
	9. Orthoptera	**	— crickets, grasshoppers, locusts
	10. Phasmida	ø	— stick insects, leaf insects
	11. Dermaptera	*	— earwigs
	12. Embioptera	ø	— —
	13. Dictyoptera	‡	— cockroaches, mantids
	14. Isoptera	‡ø	— termites
	15. Zoraptera	ø	— —
	16. Psocoptera		— psocids, booklice
Phthi-	17. Mallophaga	†	— biting lice, bird lice
raptera	18. Siphunculata	†	— sucking lice
	19. Hemiptera	**	— cicadas, leafhoppers, scale insects, aphids, plant bugs, etc.
	20. Thysanoptera	*	— thrips

division B. HOLOMETABOLA (= ENDOPTERYGOTA)

Metamorphosis complete. Pupal stage present. Wings develop internally in pupal stage.

	Order		Common names
order 21.	Neuroptera	*	— lacewings, ant lions
22.	Coleoptera	**‡	— beetles
23.	Strepsiptera		— stylopids
24.	Mecoptera		— scorpion flies
25.	Siphonaptera	†	— fleas
26.	Diptera	**†‡	— flies
27.	Lepidoptera	**‡	— butterflies, moths
28.	Trichoptera		— caddis flies
29.	Hymenoptera	**	— sawflies, ants, bees, wasps

* — orders of limited economic importance in relation to cultivated plants.

** — orders of major economic importance in relation to cultivated plants.

† — orders of importance as parasites of vertebrate animals.

‡ — orders important as pests of timber, of stored foods, or of public health significance.

ø — orders absent from Great Britain.

The orders of insects

In this section the main features characteristic of each insect order are enumerated together with an indication of the economic importance of the order, the approximate number of described species for the world as a whole and the number of species known to occur in Great Britain.

Also, for those orders of economic significance the main species of practical importance present in Great Britain are listed together with their food sources (eg, plants attacked in the case of plant pests). For further information on individual pest species the reader is referred to the excellent series of advisory leaflets published by the Ministry of Agriculture, Fisheries and Food and the applied entomological tests listed on page 238.

ORDER 1. THYSANURA — three-pronged bristle tails, silverfish

(about 500 species world wide — 9 in Great Britain)

Primitive wingless insects showing no metamorphosis. Abdomen 11-segmented and bearing pairs of small appendages on some segments. Antennae long and thread-like with many joints. Biting mouthparts. Three long tail filaments.

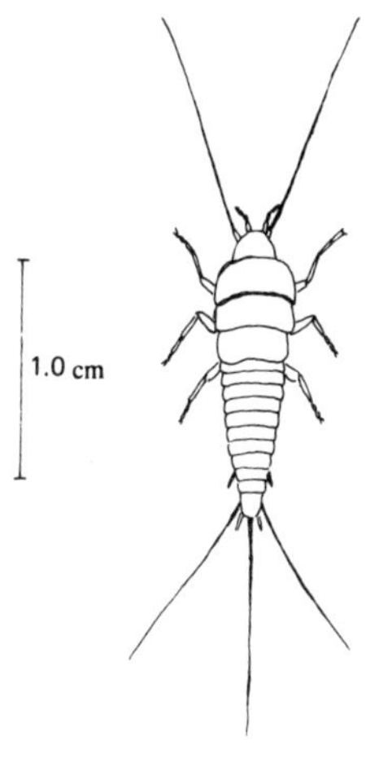

a silverfish

Some species, eg, the silverfish, *Lepisma saccharina*, are minor scavenging pests in kitchens and food premises and also may damage books and papers in older buildings. A few species have been recorded as minor pests of growing plants.

Main species of economic importance in Great Britain

Common name	*Scientific name*	*Principal hosts*	*Remarks*
firebrat	*Thermobia domestica*	nuisance pests in buildings	restricted to warm premises
silverfish	*Lepisma saccharina*	nuisance pests in buildings	restricted to warm premises

ORDER 2. DIPLURA — two-pronged bristle tails

(about 600 species world wide — 12 in Great Britain)

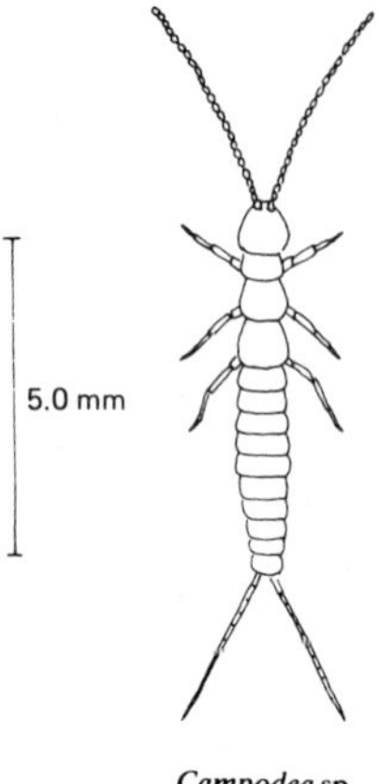

Campodea sp.

Primitive wingless insects showing no metamorphosis. Abdomen 11-segmented with most segments bearing a pair of small lateral appendages. Antennae many jointed. Biting mouthparts enclosed in the lower part of the head. A pair of long tail filaments or pincers. Eyes absent. Malpighian tubules vestigial or absent.

Mostly small whitish insects widely distributed in moist concealed situations, eg, under stones, in leaf mould. Of no economic significance.

ORDER 3. PROTURA

(about 170 species world wide — 12 in Great Britain)

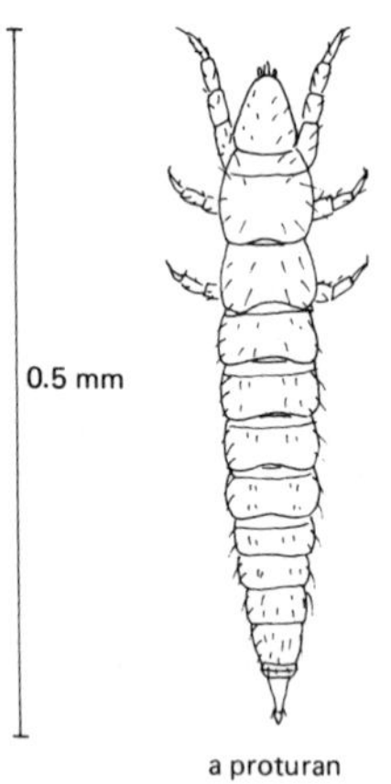

a proturan

Minute (< 2.0 mm) primitive wingless insects showing little metamorphosis. Abdomen 11-segmented with a pair of small appendages on first three segments. No antennae or eyes. Piercing enclosed mouthparts. Tail filaments absent. Malpighian tubules present as small papillae.

Extremely small whitish insects found in concealed situations (under stones, logs, etc) and in peaty soils and leaf litter. Of no economic significance. The Protura are not included among the insects by some authors because of the absence of antennae.

ORDER 4. COLLEMBOLA — springtails

(about 2,000 species world wide — about 300* in Great Britain)

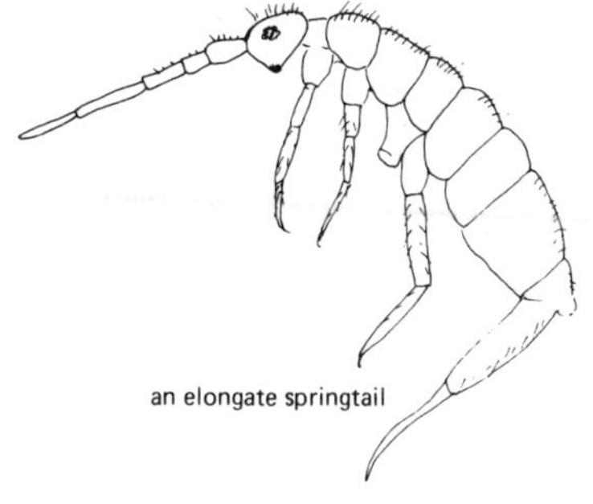

an elongate springtail

3.0 mm

***Small** (< 5 mm) **primitive wingless insects** showing little or no metamorphosis. Abdomen with only six segments, usually bearing an outgrowth (ventral tube) on segment 1 and a forked springing organ on segment 4. Antennae 4-5 jointed. Biting mouthparts enclosed in the lower part of the head. Compound eyes absent. Tracheal system usually absent. No Malpighian tubules.*

There are two sub-groups of springtails, the elongate type (Arthropleona), which are of normal elongate insect shape with the thoracic and abdominal segments distinctly separated, and the globular type (Symphypleona) with the thorax and abdomen joined into a single globular mass in which the segments cannot be distinguished.

Springtails are very common on and in the surface layers of the soil and in leaf mould under moist conditions. Most species are saprophagous and do not attack living plants. Others however may attack seedlings of various plants shortly after germination at about the time of emergence through the soil, resulting in poor plant establishment. As springtails respire through their general body surface most species are confined to moist situations, but a few feed openly on the leaves of plants, in particular the lucerne flea (*Sminthurus viridis*) which can be a serious pest of lucerne and clovers.

Main species of economic importance in Great Britain

Common name	*Scientific name*	*Principal hosts* †	*Remarks*
garden springtail	*Bourletiella hortensis*‡	seedlings of various plants	
gun powder mites	*Hypogastrura armata* and other spp.	mushrooms	eat small dry pits in caps and stalks
lucerne flea	*Sminthurus viridis*	lucerne, clovers	eats "windows" in leaves
white blind springtails	*Onychiurus* spp.	seedlings of many crop and garden plants	

* — see footnote at bottom of page 75.

† — plants attacked.

‡ — other species in addition to *B. hortensis* are probably involved.

ORDER 5. EPHEMEROPTERA — mayflies

(about 1,200 species world wide—46 in Great Britain)

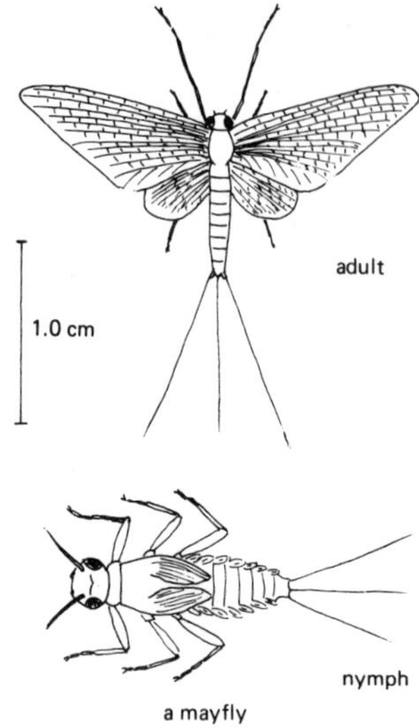

a mayfly

Delicate soft-bodied insects with two pairs of membranous wings, the hindwings being much smaller than the forewings. Numerous veins and cross veins. Abdomen with three long tail filaments. Antennae short. Mouthparts not developed. Two adult stages occur, a sub-imago (dun) which is not sexually mature, and an imago (spinner) which is sexually mature. Development hemimetabolous. Immature stages (nymphs) are aquatic in flowing fresh water. Nymphs normally possess three long tail filaments and membranous external gills on the abdomen. There are many (30-40) moults.

Mayfly nymphs are common in swift flowing streams and rivers where they feed on microscopic plant life (algae, etc) which grow on the surface of submerged stones and other objects. Most species are flattened and adapted for clinging and crawling over the surface of stones in fast water, but some are free swimming. Development takes at least a year for most species but the adults live for only a day or two and do not feed. Mayfly nymphs and adults are an important item in the diet of many freshwater fish but otherwise are of no economic significance.

ORDER 6. ODONATA — dragonflies

(about 5,000 species world wide — about 42 in Great Britain)

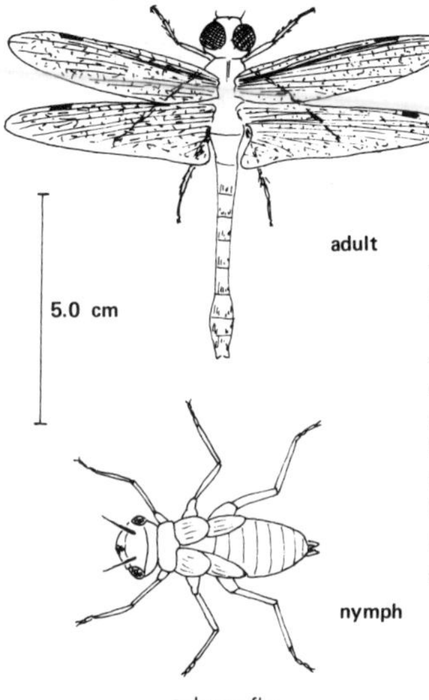

a dragonfly

Large insects with two pairs of similar membranous wings, each with a large number of veins and cross veins and a prominent stigma. Abdomen extremely long and slender. Antennae very short. Eyes extremely large and prominent. Biting mouthparts. Development hemimetabolous. Immature stages (nymphs) aquatic in still fresh water. Mouthparts of nymphs placed at the end of an extendable arm (mask). Gills in the form of tail filaments or internally in rectum.

Dragonflies are predatory both as nymphs and as adults. The prey of the adults consists mostly of other insects but besides aquatic insects, the nymphs may feed on worms, tadpoles, and even small fish. The larger species, which may be up to 200 mm in wing span, are strong and rapid fliers but the smaller damselflies are slow and feeble in flight. Other than the slight benefit derived from their predatory habit and the fact that nymphs provide food for some freshwater fish, the dragonflies are of no economic significance.

ORDER 7. PLECOPTERA — stoneflies

(about 3,000 species world wide — 34 in Great Britain)

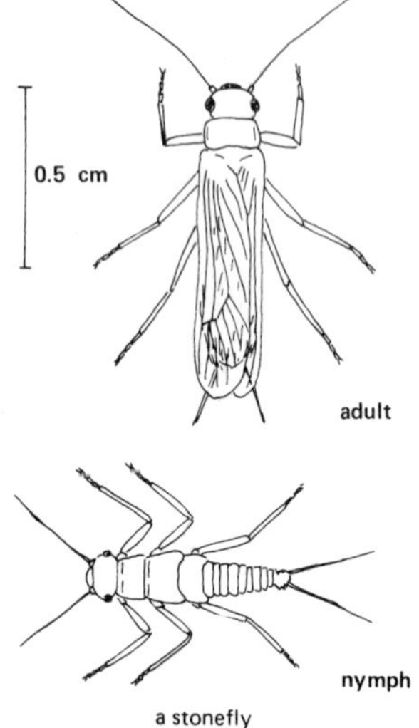

a stonefly

Moderate sized soft-bodied insects with two pairs of membranous wings which are folded flat over the back at rest. Numerous veins and cross veins. Abdomen usually with two tail filaments. Antennae of moderate length. Mouthparts of biting pattern but weakly developed. Development hemimetabolous. Immature stages (nymphs) are aquatic in flowing fresh water with a few species terrestrial in moist places. Nymphs possess two prominent tail filaments. External abdominal gills may be present.

Stonefly nymphs are common in freshwater streams particularly in the colder, more turbulent headwaters. Stonefly nymphs (like mayfly nymphs with which they may be confused) feed mostly on minute plant life on the substrate, but some species are predatory. The adults, which are drab in colour, are feeble fliers and are rarely found far from water. Apart from contributing to the diet of some freshwater fish, stoneflies are of no economic significance.

ORDER 8. GRYLLOBLATTODEA

(16 species world wide — none in Great Britain)

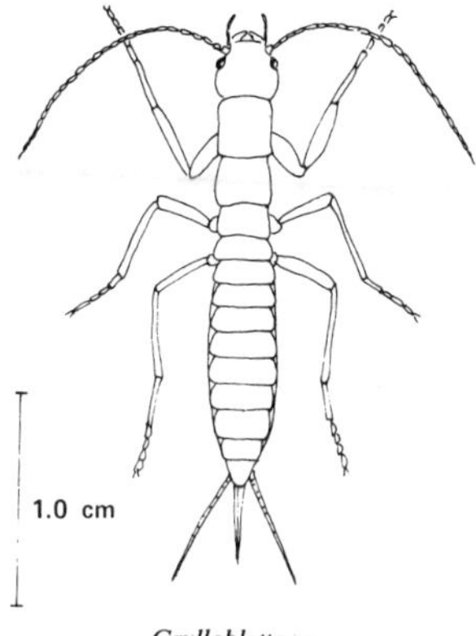

Grylloblatta sp.

Moderate sized wingless insects. Eyes reduced or absent. Abdomen with 8-jointed cerci. Moderately long antennae with many similar joints. Mouthparts of biting pattern. Legs similar to each other. Tarsi 5-jointed. Prominent ovipositor present in the female. Metamorphosis hemimetabolous.

The Grylloblattodea are a very small group of rather primitive insects of entomological interest mainly because they are intermediate between the cockroaches (Dictyoptera) and crickets (Orthoptera). They are not known to occur in the Southern hemisphere and are of no economic significance.

ORDER 9. ORTHOPTERA — crickets, grasshoppers, locusts

(about 20,000 species world wide—30 in Great Britain)

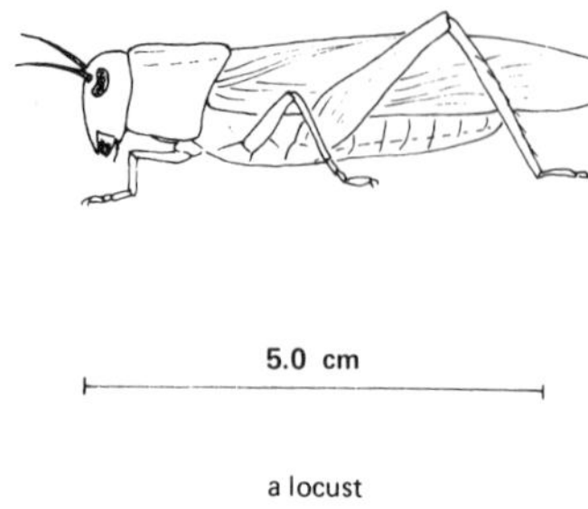

a locust

Medium-sized to large insects with well-developed exoskeleton. Usually with two pairs of wings and the forewings rather heavier in texture than the hindwings. Some species with reduced wings or wingless. Hind legs usually enlarged for jumping. Cerci short and unjointed. Antennae long and filamentous (as in crickets) or short (as in locusts). Mouthparts of generalised biting pattern. Prominent ovipositor usually present in female. Special sound producing and receiving organs often present. Development hemimetabolous.

In older entomological texts the cockroaches, mantids and stick insects were often included in the Orthoptera but these groups are now usually placed in separate orders so that the Orthoptera proper comprises only the crickets, grasshoppers, locusts, and close relatives. Most Orthoptera are rather general plant feeders though there is perhaps an overall preference for graminaceous plants (grasses and cereals). All members of the Orthoptera possess mouthparts of a generalised biting type with powerful biting mandibles and most species therefore can tackle the toughest of plants.

The most serious pests within the Orthoptera are the locusts which are large tropical or subtropical grasshoppers that develop in enormous numbers. There are several species and each occurs in two distinct phases which differ in appearance and behaviour. The **solitary phase**, which develops from sparse populations, does not band together or migrate and so is relatively harmless. The **gregarious phase** however, which is produced when overcrowding occurs in breeding areas, gathers into huge swarms which migrate over long distances, and is extremely destructive.

Main species of economic importance in Great Britain

Common name	*Scientific name*	*Principal hosts*	*Remarks*
house cricket	*Acheta domesticus*	occasionally damages root crops	nuisance pest in buildings

ORDER 10. PHASMIDA — stick insects, leaf insects

(about 2,500 species world wide — none in Great Britain)

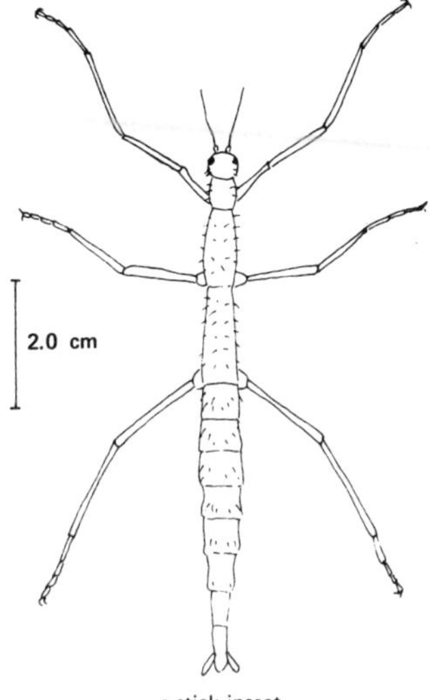

a stick insect

Mostly large insects of elongate cylindrical form (stick insects) or flattened (leaf insects). Winged or wingless. Forewings when present usually small. Middle and hind segments of thorax in stick insects very long. Legs similar to each other. Cerci short and unjointed. Biting mouthparts. Development hemimetabolous.

The stick insects and leaf insects are mostly of interest because of their remarkable resemblance to shoots and leaves of plants for purposes of camouflage. All are plant feeders and some species of stick insects are of significance as plant pests in a few tropical areas. Several species are reared as laboratory insects in Great Britain.

ORDER 11. DERMAPTERA — earwigs

(about 1,200 species world wide — 4 in Great Britain)

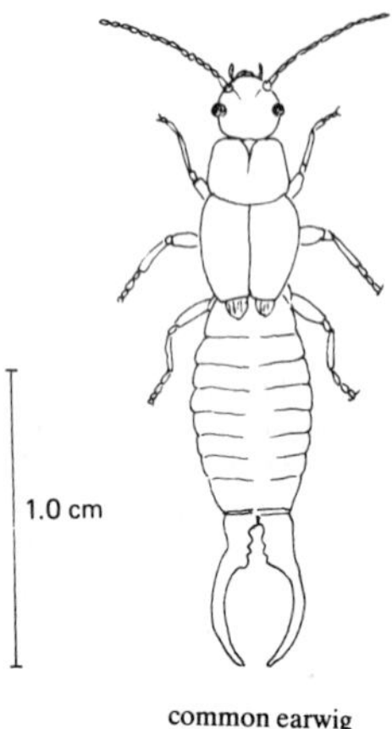

common earwig

*Small to moderate sized, rather flattened insects. Forewings reduced to small leathery flaps (***tegmina***). Hindwings membranous and semi-circular; folded under forewings at rest. Many species wingless. Cerci unjointed and modified into prominent horny forceps. Mouthparts of generalised biting type. Development hemimetabolous.*

Most earwigs are nocturnal and omnivorous in habit. A few species regularly injure plants by biting into developing flower buds and fruits in particular. Some species on the other hand are predatory and a very few are parasitic on other insects.

Main species of economic importance in Great Britain

Common name	*Scientific name*	*Principal hosts*	*Remarks*
common earwig	*Forficula auricularia*	chrysanthemum, dahlia, fruits	may damage flowers and fruits, nocturnal; also predatory and thus may be beneficial

ORDER 12. EMBIOPTERA

(at least 200 species world wide — none in Great Britain)

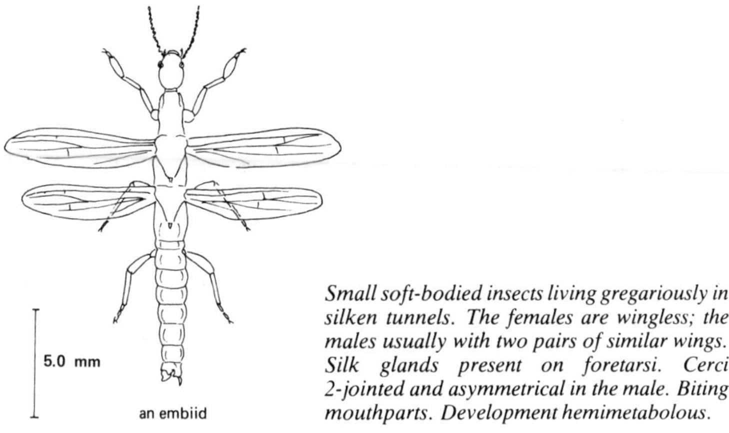

Small soft-bodied insects living gregariously in silken tunnels. The females are wingless; the males usually with two pairs of similar wings. Silk glands present on foretarsi. Cerci 2-jointed and asymmetrical in the male. Biting mouthparts. Development hemimetabolous.

The Embioptera are sometimes referred to as web-spinners due to the fact that they build silken nests in which they live gregariously. They are essentially tropical insects. All species feed on plant material, living or dead, but are rarely of economic significance.

ORDER 13. DICTYOPTERA — cockroaches, mantids

(about 6,000 species world wide — about 10 in Great Britain*)

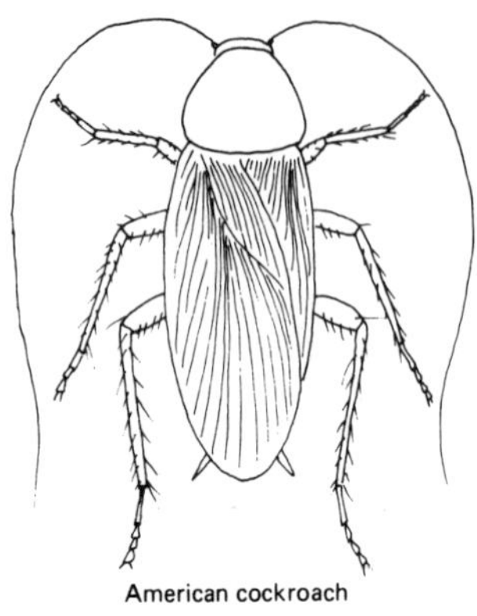
American cockroach

1.0 cm

Medium-sized to large insects with well developed exoskeleton. Usually with two pairs of wings. Forewings rather heavier in texture than hindwings. Some species wingless. Forelegs modified for clasping in the sub-order Mantodea (mantids). Antennae long with numerous small joints. Mouthparts of generalised biting pattern. Eggs produced in an ootheca. Development hemimetabolous.

The cockroaches and mantids are sometimes placed in two separate orders, the Blattaria and the Mantodea, respectively, but are here treated as sub-orders of the single order Dictyoptera.

The cockroaches, of which there are about 3,500 species world wide are omnivorous scavengers consuming human food of all kinds as well as decaying organic matter of diverse origin. Several species are pests of major public health importance because of their association with human habitation. Large populations often occur in city sewage systems besides being resident in buildings wherever food is available. Some species occasionally attack growing plants.

The mantids on the other hand are entirely predatory on insects and other small arthropods. They are therefore regarded as beneficial though they never seem to occur in large numbers.

Main species of economic importance in Great Britain

Common name	*Scientific name*	*Principal hosts*	*Remarks*
American cockroach	*Periplaneta americana*	organic matter of all kinds	large species; only in well heated premises; common on ships
common cockroach	*Blatta orientalis*		have been reported occasionally as damaging to glasshouse plants.
German cockroach	*Blatella germanica*		

* Cockroaches are essentially tropical/sub-tropical insects and only three species (confined to the South) are indigenous to Great Britain. However, several introduced species (listed above) are serious pests and other species often occur on ships.

ORDER 14. ISOPTERA — termites.

(about 2,000 species world wide — none in Great Britain)

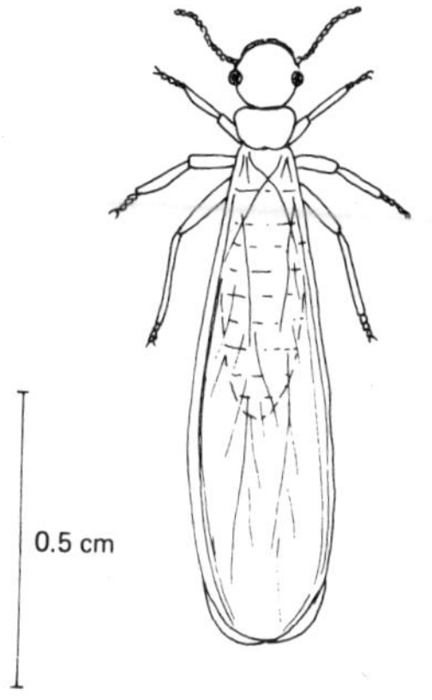

a termite (winged sexual form)

Moderate sized, rather soft-bodied insects living in large colonies composed of few reproductive forms and large numbers of wingless sterile workers and soldiers. Two pairs of very similar wings possessed by reproductives but shed after the mating flight. Antennae of moderate length with a considerable number of small similar joints. Mouthparts of generalised biting pattern. Cerci short. Development hemimetabolous with only slight metamorphosis.

Termites are sometimes referred to as white ants. This term is unfortunate as they are not related though they do show many parallel features of social organisation. All termite species live in complex colonies on which they are utterly dependent. Other than for the brief mating flight of the sexual forms all individuals spend their entire life in darkness within the nest or associated structures.

Termites feed mostly on dead plant material and some are serious pests of timber and buildings and other wooden structures in tropical countries. However, a few species attack growing plants and achieve pest status for that reason. In other situations termites have a beneficial role in assisting with soil formation in the drier parts of the tropics.

ORDER 15. ZORAPTERA

(22 species world wide — none in Great Britain)

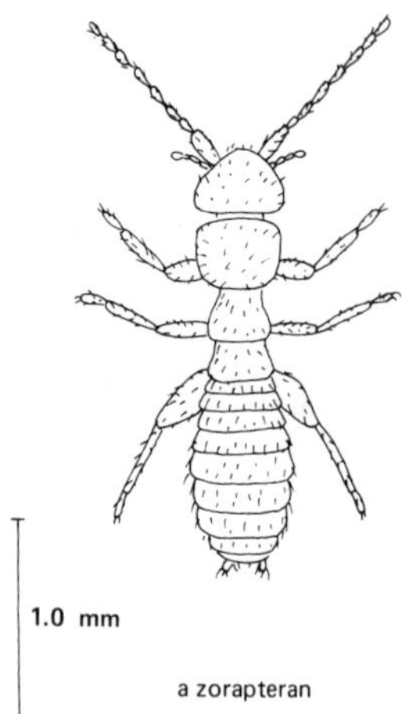

a zorapteran

Small, soft-bodied, usually wingless insects. Wings when present membranous and with few veins. Cerci small and unjointed. Generalised biting mouthparts. Development hemimetabolous with only slight metamorphosis.

The Zoraptera are a very small order of insects of uncertain affinity and mostly tropical in distribution. They are cryptic in habit and so far as is known fungal feeding. They are of no economic significance.

ORDER 16. PSOCOPTERA — psocids, booklice

(about 1,700 species world wide — about 70* in Great Britain)

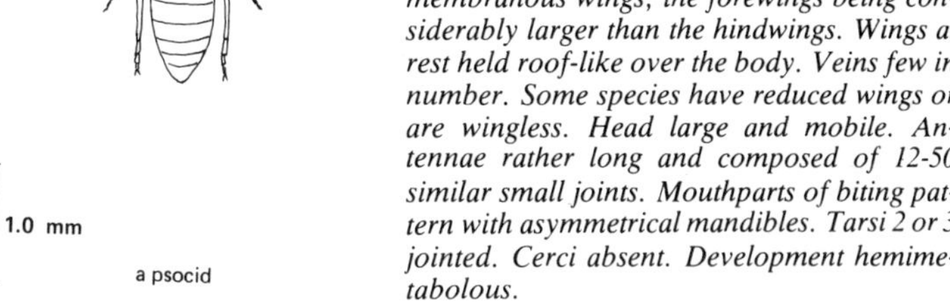

a psocid

Small soft-bodied insects with two pairs of membranous wings, the forewings being considerably larger than the hindwings. Wings at rest held roof-like over the body. Veins few in number. Some species have reduced wings or are wingless. Head large and mobile. Antennae rather long and composed of 12-50 similar small joints. Mouthparts of biting pattern with asymmetrical mandibles. Tarsi 2 or 3 jointed. Cerci absent. Development hemimetabolous.

Psocids are essentially scavengers feeding on fragments of organic matter of plant or animal origin. Feeding in most cases is probably more on the micro-organisms associated with such materials, especially fungi, rather than on the substrates themselves.

Some species are minor pests of stored human foods and of other materials such as old papers and museum specimens. Their presence is normally associated with damp conditions.

Main species of economic importance in Great Britain

Common name	*Scientific name*	*Principal hosts*	*Remarks*
psocids (booklice)	*Trogium pulsatorium* *Lepinotus inquilinus* *L. patruelis* *Liposcelis* spp.	stored foods and mouldy commodities	nuisance pests in buildings especially where food is stored and the environment is damp

* For many insect orders it is impossible to give an exact figure for the number of species because of incompleteness of current knowledge. This applies particularly to the larger orders such as Coleoptera, Lepidoptera, Hymenoptera and Diptera.

ORDER 17. MALLOPHAGA — biting lice, bird lice

(about 3,000 species world wide — about 500 in Great Britain)

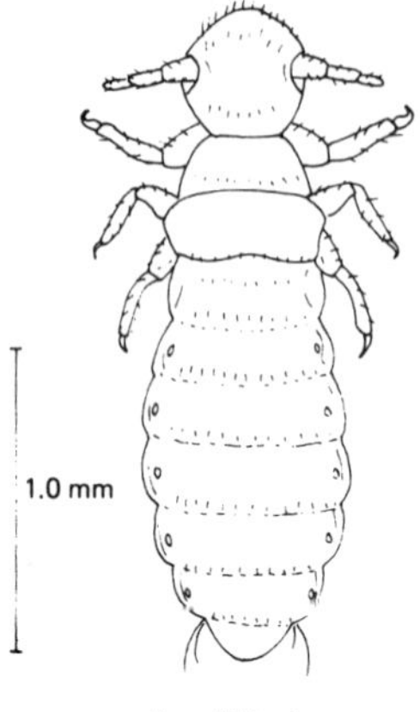

sheep biting louse

Small wingless insects permanently ectoparasitic on birds and in some cases on mammals. Heads broadly flattened. Eyes reduced. Ocelli absent. Antennae 3-5 jointed. Mouthparts of modified biting pattern. Prothorax distinct. Tarsi with 1 or 2 joints. Cerci absent. Development hemimetabolous with slight metamorphosis.

Biting lice are entirely parasitic (on birds for the most part with a few species infesting mammals). The entire life cycle is spent on the host and they are able to survive for only short periods away from the host. Feeding is mainly on feathers, hair and other epidermal products but blood meals may sometimes be taken. Low levels of infestation seem to have little effect on animals in the wild but with domesticated species biting lice can be a serious problem.

Main species of economic importance in Great Britain

Common name	*Scientific name*	*Principal hosts*	*Remarks*
cattle biting louse	*Damalinia bovis*	cattle	common but rarely important
chicken body louse	*Menacanthus stramineus*	poultry	often around vent
chicken fluff louse	*Goniocotes gallinae*	poultry	
chicken wing louse	*Lipeurus canopus*	poultry	affects neck and head
chicken shaft louse	*Menapon gallinae*	poultry	
dog biting louse	*Trichodectes canis*	dog	
sheep biting louse	*Damalinia ovis*	sheep	

ORDER 18. SIPHUNCULATA (ANOPLURA) — sucking lice

(about 300 species world wide — about 25 in Great Britain)

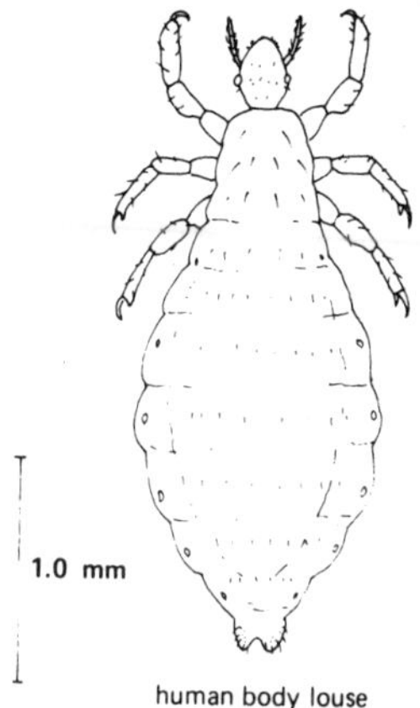

human body louse

Small wingless insects permanently ectoparasitic on mammals. Body flattened dorso-ventrally. Head rather small and pointed. Prothorax not distinct from rest of thorax. Eyes reduced or absent. Ocelli absent. Antennae 3-5 jointed. Mouthparts modified for piercing and sucking; retractable. Prominent legs with unjointed tarsi bearing large single claws. Cerci absent. Development hemimetabolous with slight metamorphosis.

The sucking lice are entirely blood feeding and are permanently parasitic on mammals. They can survive for only short periods away from the host. Two species infest man. *Pediculus humanus*, the human louse, exists in two forms, the head louse and the body louse. It is important not only for the debilitating and irritant effects if produces but also because it acts as a vector for several serious diseases, in particular typhus. The second species is the crab louse, *Pthirus pubis* which infests mainly the pubic region. It is not disease carrying. Other species of lice are important pests of livestock.

Main species of economic importance in Great Britain

Common name	*Scientific name*	*Principal hosts*	*Remarks*
blue cattle louse	*Solenopotes capillatus*	cattle	
body louse	*Pediculus humanus corporis*	man	in clothing
crab louse	*Pthirus pubis*	man	confined to the pubic region
dog sucking louse	*Linognathus setosus*	dog	
head louse	*Pediculus humanus capitis*	man	on the head
horse sucking louse	*Haematopinus asini*	horse	
long-nosed cattle louse	*Linognathus vituli*	cattle	
pig louse	*Haematopinus suis*	pig	common
sheep foot louse	*Linognathus pedalis*	sheep	confined to legs and feet
sheep-sucking louse	*L. ovillus*	sheep	common
short-nosed cattle louse	*Haematopinus eurysternus*	cattle	common

ORDER 19. HEMIPTERA — cicadas, leafhoppers, scale insects, aphids, plant bugs, etc.

(about 65,000 species world wide – about 1,650 in Great Britain)*

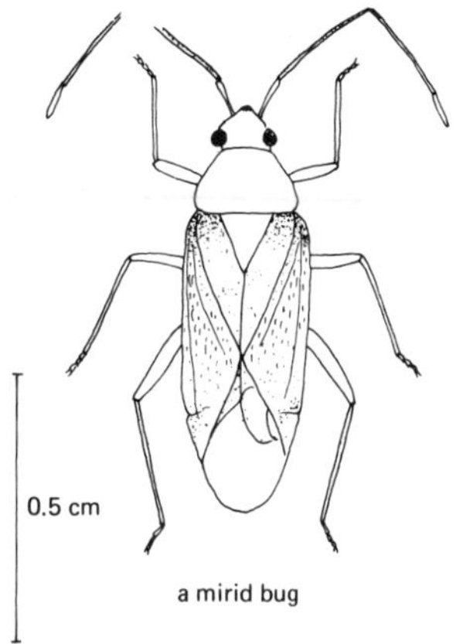

a mirid bug

Small to large insects usually with two pairs of wings but apterous forms common. Wings when present with the forewings larger and of heavier texture than hindwings; uniformly so in the sub-order Homoptera; with the tip more membranous than the base in the sub-order Heteroptera. Mouthparts of piercing-sucking pattern and lacking palps. Development hemimetabolous with marked metamorphosis in some groups but little in others.

The Hemiptera are a large group of apparently diverse insects but all possess the common feature of specialised piercing/sucking mouthparts. Most are plant feeding and the order includes some of the most important groups of plant pests in the world, such as leafhoppers, aphids and scale insects. Besides the effects resulting from removal of plant sap, many Hemiptera injure plants as a result of toxic effects from their saliva, causing distortion and stunting of growth. All plant feeding species produce honeydew on which sooty moulds often develop causing another problem. In addition, many species (particularly aphids and leafhoppers) are major vectors of plant virus diseases.

Some species of Hemiptera are predatory on other insects and may thus be beneficial while a few, including the human bed bug, are blood sucking on higher animals.

* — See footnote at bottom of page 75

Main species of economic importance in Great Britain

Common name	*Scientific name*	*Principal hosts*	*Remarks*
Pests of plants			
sub-order Heteroptera			
Lygaeidae (plant bugs)			
apple capsid	*Plesiocoris rugicollis*	apple, currant, gooseberry	rare in sprayed orchards
common green capsid	*Lygocoris pabulinus*	apple, pear, berry fruits, many perennial weeds	scars and disfigures fruit and foliage
potato capsid	*Calocoris norvegicus*	potato	produces small red spots on leaves which develop to holes; a minor pest, mostly on headlands
tarnished plant bug	*Lygus rugulipennis*	many plants	yellow/white speckling of leaves later turning brown
sub-order Homoptera			
Cicadellidae (leaf hoppers)			
fruit tree leafhoppers	*Alnetoidea alneti* *Edwardsiana crataegi*	apple	produce white flecks on leaves
glasshouse leafhopper	*Hauptidia maroccana*	cucumber, tomato, chrysanthemum and many other ornamental plants	a problem where integrated control programmes are being used
rose leafhopper	*Edwardsiana rosae*	rose	produces white flecks on leaves
Tingidae (lace bugs)			
rhododendron bug	*Stephanitis rhododendri*	rhododendron	leaves become speckled then bleached
Psyllidae (suckers)			
apple sucker	*Psylla mali*	apple	rare in sprayed orchards
pear sucker	*P. pyricola*	pear	damages developing blossom and leaf trusses
Aphids			
apple grass aphid	*Rhopalosiphum insertum*	grasses	causes slight leaf curling on apples
auricula root aphid	*Pemphigus auriculae*	primula	found on roots
beech aphid	*Phyllaphis fagi*	beech	
bird-cherry aphid	*Rhopalosiphum padi*	grasses, wheat, oats, barley	vector of BYDV

Common name	*Scientific name*	*Principal hosts*	*Remarks*
Aphids — continued			
black bean aphid	*Aphis fabae*	field, broad, french and runner beans, beet	overwintering eggs on spindle
blackberry-cereal aphid	*Sitobion fragariae*	cereals, grasses	mostly on ears
blackcurrant aphid	*Cryptomyzus galeopsidis*	blackcurrant	
bulb and potato aphid	*Rhopalosiphonius latysiphon*	potato stolons	
cabbage aphid	*Brevicoryne brassicae*	cabbage, cauliflower, sprouts and other brassicas	mealy grey in colour
cereal leaf aphid	*Rhopalosiphum maidis*	maize, barley	
cherry blackfly	*Myzus cerasi*	cherry	causes leaf curl and stunting of shoots
chrysanthemum aphid	*Macrosiphoniella sanborni*	chrysanthemum	forms colonies principally on stem below the head
currant-lettuce aphid	*Nasonovia ribisnigri*	gooseberry, lettuce	eggs on gooseberry over winter
currant root aphid	*Eriosoma ulmi*	currants, gooseberry	attacks roots
currant sowthistle aphid	*Hyperomyzus lactucae*	blackcurrant	curls leaves
damson-hop aphid	*Phorodon humuli*	plum, damson, hops	attacks primarily tips of shoots; infests cones
fern aphid	*Idiopterus nephrelepidus*	ferns	
fescue aphid	*Metopolophium festucae*	cereals, grasses	entire life cycle spent on grasses and cereals
glasshouse and potato aphid	*Aulacorthum solani*	sprouting potatoes, lettuce	
gooseberry aphid	*Aphis grossulariae*	gooseberry, currants	
grain aphid	*Sitobion avenae*	cereals, grasses	entire life cycle spent on cereals and grasses; infests heads of cereals, particularly wheat
green apple aphid	*Aphis pomi*	apple, pear, quince, hawthorn	some leaf curl; can seriously affect young trees
green spruce aphid	*Elatobium abietinum*	spruce	
leaf-curling plum aphid	*Brachycaudus helichrysi*	plum, damson	attacked leaves become curled
lettuce root aphid	*Pemphigus bursarius*	lettuce	attacks roots; affected plants wilt
mealy plum aphid	*Hyalopterus pruni*	plum, damson	attacks primarily tips of shoots

Common name	*Scientific name*	*Principal hosts*	*Remarks*
Aphids — (continued)			
melon and cotton aphid	*Aphis gossypii*	cucumber, chrysanthemum	common under glass; causes extensive distortion of cucumber
mottled arum aphid	*Aulacorthum circumflexum*	arum, cyclamen and many other ornamental plants	primarily under glass
orchid aphid	*Cerataphis orchidearum*	orchids, palms	scale-like in appearance
pea aphid	*Acyrthosiphon pisum*	pea, beans, clover	infests flowers, causes pod distortion in peas
peach aphid	*Brachycaudus schwartzi*	peach	
peach-potato aphid	*Myzus persicae*	peach, potato, sugarbeet, and many other plants	vector of many plant viruses
pear-bedstraw aphid	*Dysaphis pyri*	pear	
potato aphid	*Macrosiphum euphorbiae*	potato	causes false leaf roll symptoms
redcurrant blister aphid	*Cryptomyzus ribis*	red, white and blackcurrant	causes blisters on leaves
rose aphid	*Macrosiphum rosae*	rose	
rose-grain aphid	*Metopolophium dirhodum*	wheat, grasses	particularly on lower leaves
rose leaf curling aphid	*Dysaphis devecta*	apple	reddish distorted leaves
rosy apple aphid	*D. plantaginea*	apple	severe leaf distortion; deformed fruit
rubus aphid	*Amphorophora rubi*	raspberry, loganberry, blackberry	
shallot aphid	*Myzus ascalonicus*	shallot, strawberry	can cause extensive distortion of crown in strawberry
strawberry aphid	*Chaetosiphon fragaefolii*	strawberry	virus vector
tulip bulb aphid	*Dysaphis tulipae*	tulip, iris, gladiolus	attacks sprouting corms
willow-carrot aphid	*Cavariella aegopodii*	carrot, parsnip, celery, parsley	serious but sporadic pest of carrot
woolly aphid	*Eriosoma lanigerum*	apple, cotoneaster	produces dense cotton wool-like strands on woody parts
Adelgids			
Douglas fir adelges	*Adelges cooleyi*	Douglas fir	
spruce pineapple gall adelges	*A. abietis, A. viridis*	spruce, larch	causes "pineapple" galls on new shoots
Scale insects			
beech scale	*Cryptococcus fagisuga*	beech	

Common name	Scientific name	Principal hosts	Remarks
Scale insects — (continued)			
brown scale	*Parthenolecanium corni*	many woody plants	
camellia scale	*Hemiberlesia rapax*	camellia, begonia, cacti, orchids	
cattleya scale	*Parlatoria proteus*	palms, orchids	
cushion scale	*Chloropulvinaria floccifera*	camellia, orchids	
fern scale	*Pinnaspis aspidistrae*	ferns	
hemispherical scale	*Saissetia coffeae*	wide range of glasshouse plants	
juniper scale	*Carulaspis* sp.	juniper	
mussel scale	*Lepidosaphes ulmi*	apple	uncommon in sprayed commercial orchards
oleander scale	*Aspidiotus nerii*	palms and other ornamental plants	
orchid scale	*Diaspis boisduvalii*	palms, orchids	
oyster scale	*Quadraspidiotus ostreaeformis*	apple, pear, plum, birch	
rose scale	*Aulacaspis rosae*	rose	serious pest of roses under glass
soft scale	*Coccus hesperidum*	wide range of glasshouse plants	
woolly vine scale	*Pulvinaria vitis*	vines	
Mealybugs			
citrus mealybug	*Planococcus citri*	wide range of glasshouse plants	difficult to control as protected by waxy secretions. Subterranean mealy-bugs attack roots
glasshouse mealy bug	*Pseudococcus obscurus*		
long-tailed mealy bug	*P. adonidum*		
subterranean mealybugs	*Rhizoecus* spp.		
Whiteflies			
glasshouse whitefly	*Trialeurodes vaporariorum*	wide range of glasshouse plants	produces much honey-dew and sooty mould
rhododendron whitefly	*Dialeurodes chittendeni*	rhododen-dron	leaves become yellowed; honeydew and sooty mould develop
Predators of plant pests			
black-kneed capsid	*Blepharidopterus angulatus*	fruit tree red spider mite	
common flower bug	*Anthocoris nemorum*	aphids, scale insects, apple sucker, mites	
tree damsel bug	*Himacerus apterus*		
Pests of man and animals			
bed bug	*Cimex lectularius,*	humans, poultry	bloodsucking
pigeon bug	*C. columbarius*	poultry	bloodsucking

ORDER 20. THYSANOPTERA — thrips

(about 4,000 species world wide — about 160 in Great Britain)

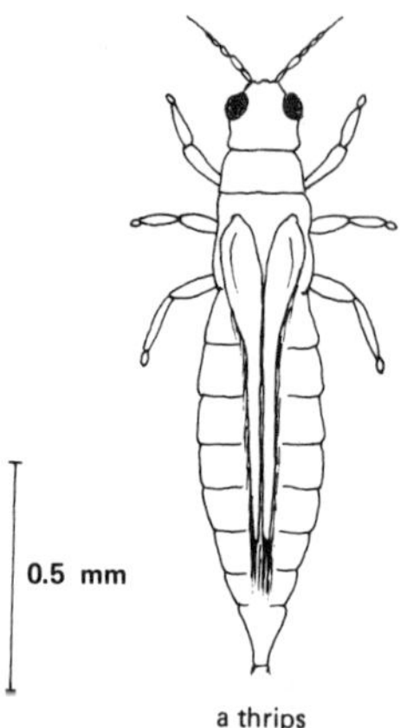

a thrips

Small slender insects with two pairs of long strap-like wings with few veins and fringed with long hairs. Some species wingless. Prominent prothorax. Legs terminate in small bladder-like structures rather than claws. Antennae short and 6-10 jointed. Mouthparts asymmetrical and adapted for piercing. Development hemimetabolous with an incipient pupal stage.

Thrips are small insects and easily overlooked. Most species are plant feeding and damage is usually seen before the insects themselves. Although thrips possess piercing mouthparts, these are very short compared to Hemiptera so that they can pierce only superficial cells of plants and suck out their contents. The result is small silvery flecks on the plant surface which eventually coalesce giving a bleached appearance. In addition there is often distortion of growth in response to the insect's injected saliva. Some species are known to transmit plant virus diseases. Parthenogenetic reproduction is common among thrips and males are rare in some species. Because the final nymphal instar is virtually a pupa (it is inactive and does not feed) the thrips provide a link between insects with incomplete metamorphosis and those that display complete metamorphosis.

Main species of economic importance in Great Britain

Common name	*Scientific name*	*Principal hosts*	*Remarks*
carnation thrips	*Thrips atratus*	carnation	
field thrips	*Thrips angusticeps*	peas, brassicas	injures very young plants
gladiolus thrips	*Thrips simplex*	gladiolus, freesia, iris	silvery flecks on leaves and flowers
glasshouse thrips	*Heliothrips haemorrhoidalis*	many glasshouse plants	common
lily thrips	*Liothrips vaneeckei*	lily	feeds on bulbs only
onion thrips	*Thrips tabaci*	onion, cabbage, chrysanthemum, cucumber, beet, and many other plants	
pea thrips	*Kakothrips pisivorus*	pea, broad bean	causes silvering of pods
pear thrips	*Taeniothrips inconsequens*	pear	may cause fruit russet
rose thrips	*Thrips fuscipennis*	rose, peach, cucumber	serious pest of roses
yellow orchid thrips	*Anaphothrips orchidaceus*	orchids	

ORDER 21. NEUROPTERA — lacewings, ant lions

(about 5,000 species world wide — 60 in Great Britain)

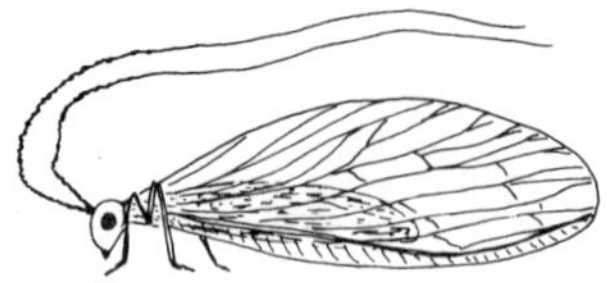

Moderate sized to rather large soft-bodied insects with two pairs of large very similar wings. Numerous veins and cross veins. Antennae rather long and thread-like. Mouthparts of biting pattern but often with very elongated pointed mandibles. Development holometabolous with campodeiform larvae. All species predatory.

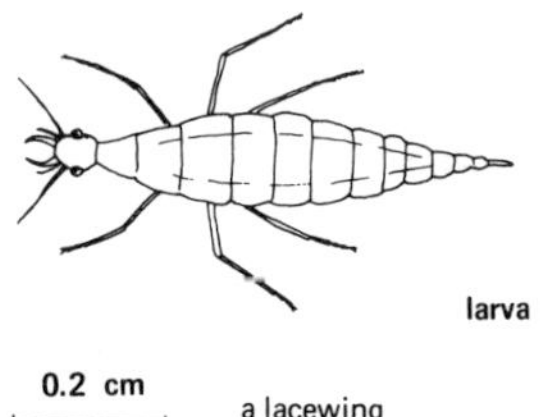

a lacewing

The Neuroptera comprise the lacewings, ant lions and their relatives. Alder flies, here included in the Neuroptera, are sometimes placed in a separate order (Megaloptera). All Neuroptera are predatory on other insects, at least in the larval stage. Where the prey consists of plant pests, as is the case with lacewings which feed mostly on aphids, their activities are beneficial.

Main species of economic importance in Great Britain

Common name	*Scientific name*	*Principal prey*	*Remarks*
brown lacewings	*Hemerobius* spp. *Kimminsia* spp.	aphids, fruit tree red spider mite	
green lacewings	*Chrysopa* spp.	aphids, suckers, leafhoppers	
powdery lacewings	*Coniopteryx* spp. *Conwentzia* spp. *Semiadalis* spp.	fruit tree red spider mite	

ORDER 22. COLEOPTERA — beetles

(about 320,000 species world wide — at least 4,000 in Great Britain)*

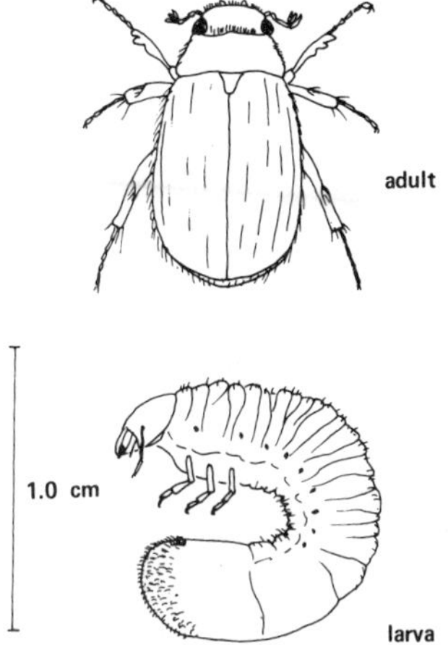

a scarab beetle

*Minute to large insects with a hard exoskeleton. Two pairs of wings; the forewings not used for flight but modified into hard horny cases (**elytra**) for the membranous hindwings. Flightless species (which retain the elytra, but not the hindwings) common. Prothorax prominent; mesothorax reduced. Mouthparts of biting pattern. Metamorphosis complete. Larvae variable in form, legless in some groups.*

The Coleoptera form the largest order of insects comprising some 40 percent of known species. They vary enormously in size and include some of the largest (up to 15 cm) as well as the smallest (about 0.5 mm) of insects. All beetles have a strongly developed hard exoskeleton except for the dorsal surface of the abdomen which is normally covered by the hindwings and elytra. Most beetles have functional wings and can fly but the flightless habit is very common.

Feeding behaviour varies enormously and almost any kind of organic matter may provide food for some species or another. The parasitic mode of life is not strongly developed but many beetles are predatory on other insects and hence beneficial when their prey consists of important plant pests such as aphids or scale insects. The ladybird beetles are particularly important in this respect. However, many species of beetles feed on living plants both as larvae and adults and the order contains some of the world's most important crop pests. Nearly every kind of cultivated plant is attacked by one or more beetles and in addition many serious pests of stored food and of timber are beetles. The list of economically important species is therefore a long one.

* — See footnote at the bottom of page 75

Main species of economic importance in Great Britain

Common name	*Scientific name*	*Principal hosts*	*Remarks*
Pests of agricultural and horticultural plants			
Curculionidae (weevils)			
apple blossom weevil	*Anthonomus pomorum*	apple	larvae feed within unopened flower buds causing "capped" blossoms; uncommon in commercial orchards
cabbage seed weevil	*Ceutorhynchus assimilis*	brassica seed crops e.g. oilseed rape	larvae feed on developing seeds
cabbage stem weevil	*C. quadridens*	brassicas	larvae tunnel in stems
clay coloured weevil	*Otiorhynchus singularis*	pip fruits, stone fruits, woody ornamentals	adults damage buds and shoots
clover leaf weevils	*Hypera postica* *H. nigrirostris*	clovers	larvae feed on flower and leaf buds
common clover weevils	*Sitona hispidulus* *S. sulcifrons*	clovers	larvae feed on root nodules
leaf weevils	*Phyllobius* spp.	fruit and ornamental trees	adults feed on foliage, larvae on roots of grasses and cereals
nut weevil	*Curculio nucum*	hazel, cobnut, filbert	larvae feed on developing nuts
pea and bean weevil	*Sitona lineatus*	peas, beans	adults notch leaves, larvae feed on root nodules
pine weevil	*Hylobius abietis*	pine	adults damage stems of young trees
rape winter stem weevil	*Ceutorhynchus picitarsis*	oilseed rape	larvae tunnel in stem near growing point
red clover seed weevils	*Apion trifolii* *A. apricans*	red clover	can cause considerable seed loss, adults damage leaves
red-legged weevil	*Otiorhynchus clavipes*	many plants	larvae feed on roots
strawberry blossom weevil	*Anthonomus rubi*	strawberry	larvae develop in flower buds
strawberry rhyncites	*Rhyncites germanicus*	strawberry, raspberry, blackberry	adults attack tips of new growth
strawberry root weevils	*Otiorhynchus rugostriatus* *O. rugifrons, O. ovatus* *Sciaphilus asperatus*	strawberry	larvae damage roots, adults notch leaves
turnip gall weevil	*Ceutorhynchus pleurostigma*	turnips and other cruciferae	larvae cause galls on roots
vine weevil	*Otiorhynchus sulcatus*	strawberry, cyclamen, nursery stock and many woody ornamentals	larvae attack roots

Common name	*Scientific name*	*Principal hosts*	*Remarks*
Pests of agricultural and horticultural plants — (continued)			
Curculionidae (weevils) *(continued)*			
white clover weevil	*Apion dichroum*	white clover	serious pest of white clover seed crops
Other plant pests			
asparagus beetle	*Crioceris asparagi*	asparagus	strips foliage
bean beetle	*Bruchus rufimanus*	peas and beans for seed	larvae feed in developing seed
blossom beetles	*Meligethus* spp.	cruciferae	damage buds and flowers of brassica seed crops
cabbage stem flea beetle	*Psylliodes chrysocephala*	brassicas	larvae tunnel in leaf stalks and stems in winter
cereal leaf beetle	*Oulema melanopa*	barley, oats, wheat	seldom serious
chafer grubs	larvae of *Melolontha melolontha* *Phyllopertha horticola* *Amphimallon solstitialis*	pasture, turf, potatoes, and many ornamental plants	larvae feed on roots
colorado beetle	*Leptinotarsa decemlineata*	potatoes	very serious pest on continent. Not yet established in UK
common click beetles (wireworms)	*Agriotes lineatus* *A. obscurus* *A. sputator* *Athous haemorrhoidalis* and other spp.	many plants, particularly cereals and potatoes	larvae feed on roots and other underground parts
flea beetles	*Phyllotreta* spp. *Chaetocnema concinna* and other spp.	brassicas	adults eat small holes in leaves and stems of seedlings
mustard beetles	*Phaedon cochleariae* *P. amoraciae*	mustard, swede, turnip, cabbage	adults and larvae feed on foliage
pygmy mangold beetle	*Atomaria linearis*	sugar beet, mangold	attacks stem just below ground level
raspberry beetle	*Byturus tomentosus*	raspberry, loganberry, blackberry	larvae feed on fruit, adults may damage buds
strawberry seed beetle	*Harpalus rufipes*	strawberry	adults bite seed from fruits
tortoise beetles	*Cassida* spp.	sugar beet	larvae cause "shot-holing" of cotyledons
Predatory beetles			
Coccinellidae (ladybird beetles)			
eleven spot ladybird	*Coccinella undecimpunctata*	aphids	

Common name	*Scientific name*	*Principal prey/hosts*	*Remarks*
Predatory beetles – *continued*			
Coccinellidae – continued (ladybird beetles)			
fourteen spot ladybird	*Propylea quatt-uordecimpunctata*	aphids	
minute black ladybird	*Stethorus punctillum*	fruit tree red spider mite	
seven spot ladybird	*Coccinella septem-punctata*	aphids	
ten spot ladybird	*Adalia decempunctata*	aphids	
twenty-two spot ladybird	*Thea vigintiduo-punctata*	aphids	
two spot ladybird	*Adalia bipunctata*	aphids	
Carabidae			
ground beetles	*Bembidion* spp. *Carabus* spp. *Harpalus* spp. *Nebria brevicollis* *Pterostichus* spp.	soil dwelling insects	may also feed on aphids and insect eggs
Staphylinidae			
rove beetles	*Aleochara* spp. *Anotylus* spp. *Atheta* spp.	soil dwelling insects	
Pests of forest trees and timber			
black pine beetles	*Hylastes* spp.	pine	adults ring bark of young trees
common furniture beetle (larvae= woodworm)	*Anobium punctatum*	primarily softwoods	most common cause of "wormy furniture"
common timberman	*Acanthocinus aedilis*		
conifer ambrosia beetle	*Xyloterus lineatus*	spruce, larch	attacks felled timber
deathwatch beetle	*Xestobium rufovillosum*	old seasoned timber especially oak	can seriously damage old buildings
elm bark beetles	*Scolytus scolytus* *S. multistriantus*	elm	vectors of Dutch elm disease fungus
false furniture beetle	*Ernobius mollis*	softwoods	attacks bark and outer sapwood only
house longhorn beetle	*Hylotrupes bajulus*	softwoods	

Common name	*Scientific name*	*Principal hosts*	*Remarks*
Pests of forest trees and timber — (*continued*)			
larch bark beetle	*Ips cembrae*	larch	bores into shoots
pine shoot beetles	*Tomicus piniperda* *T. minor*	pine	bore into shoots
post boring beetle	*Ptilinus pectinicornis*	hardwoods especially beech	
powder-post beetles	*Lyctidae*	sapwood of softwood timbers	attacks primarily fresh timber
spruce bark beetle	*Ips typographus*		
tanbark borer	*Phymatodes testaceus*	oak	
two-toothed pine beetle	*Pityogenes bidentatus*		
wood boring weevils	*Euophryum confine* *Pentarthrum huttoni*	softwoods	associated with fungal attack
Pests of stored products			
Australian spider beetle	*Ptinus tectus*	many stored foods especially those of high protein content	primarily a scavenger
bacon beetle	*Dermestes lardarius*	bacon, ham, skins, fishmeal and other materials of animal origin	
biscuit beetle	*Stegobium paniceum*	flour, biscuits, cereal products, spices	
broad-horned flour beetle	*Gnatocerus cornutus*	flour and flour products	
cadelle beetle	*Tenebrioides mauritanicus*	flour and cereal products	
carpet beetle	*Anthrenus* spp.	woollen goods, furs dried insect specimens	
cigarette bettle	*Lasioderma serricorne*	tobacco	dies out in unheated premises in winter
confused flour beetle	*Tribolium confusum*	damaged grain and flour products	closely similar to *T. castaneum*
dried fruit beetles	*Carpophilus* spp.	dried fruits	
foreign grain beetle	*Ahasverus advena*	wheat, maize meal, flour products	most frequently found on mouldy produce
fur beetle	*Attagenus pellio*	furs, skins, woollen goods	
grain weevil	*Sitophilus granarius*	wheat and other cereal grains	serious pest of stored grain
greater rice weevil	*S. zeamais*	rice, maize and other cereal grains	does not survive winter in unheated premises

Common name	*Scientific name*	*Principal hosts*	*Remarks*
Pests of stored products — *(continued)*			
khapra beetle	*Trogoderma granarium*	grain and flour products	needs relatively high temperature for reproduction, larvae diapause at low temperatures
leather beetle	*Dermestes maculatus*	skins, hides, and other materials of animal origin	
lesser grain borer	*Rhyzopertha dominica*	grain	very rarely establishes, requires source of warmth, dies out in winter
lesser mealworms	*Alphitobius diaperinus*	broken grain, flour products	common in poultry houses
lesser rice weevil	*Sitophilus oryzae*	rice, maize and other cereal grains	does not survive winter in unheated premises
meal worm	*Tenebrio molitor*	cereals and cereal products	long life cycle
merchant grain beetle	*Oryzaephilus mercator*	oilseeds, cereals and their products	
rust-red flour beetle	*Tribolium castaneum*	damaged grain and flour products	closely similar to *T. confusum* but prefers higher temperatures; only recently found in British flour mills
rust-red grain beetle	*Cryptolestes ferrugineus*	grain and flour products	
saw-toothed grain beetle	*Oryzaephilus surinamensis*	grain and cereal products	the most important pest of farm stored grain in Great Britain
yellow mealworm	*Tenebrio molitor*	whole grain. meal	long life cycle

ORDER 23. STREPSIPTERA — stylopids

(about 300 species world wide — 8 in Great Britain)

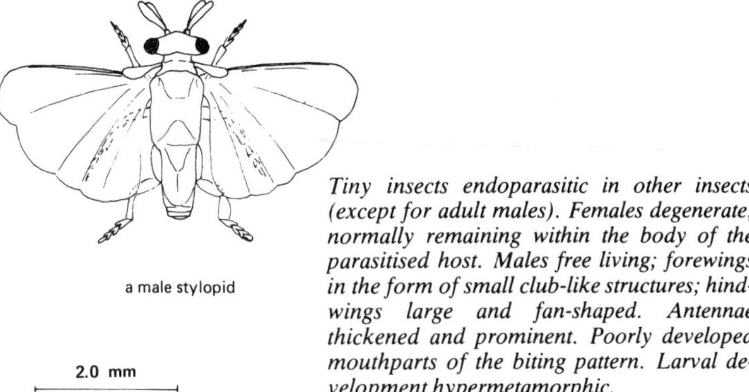

a male stylopid

2.0 mm

Tiny insects endoparasitic in other insects (except for adult males). Females degenerate, normally remaining within the body of the parasitised host. Males free living; forewings in the form of small club-like structures; hind-wings large and fan-shaped. Antennae thickened and prominent. Poorly developed mouthparts of the biting pattern. Larval development hypermetamorphic.

The stylopids are a small group of insects related to the Coleoptera and chiefly of interest because of their aberrant structure and biology. Insects which are parasitised by them are reduced in vigour and often rendered sterile but are not killed, in contrast to the parasitic Hymenoptera and Diptera. The most common hosts of stylopids are bees and other social Hymenoptera. The order is of little economic significance.

ORDER 24. MECOPTERA — scorpion flies

(about 400 species world wide — 4 in Great Britain)

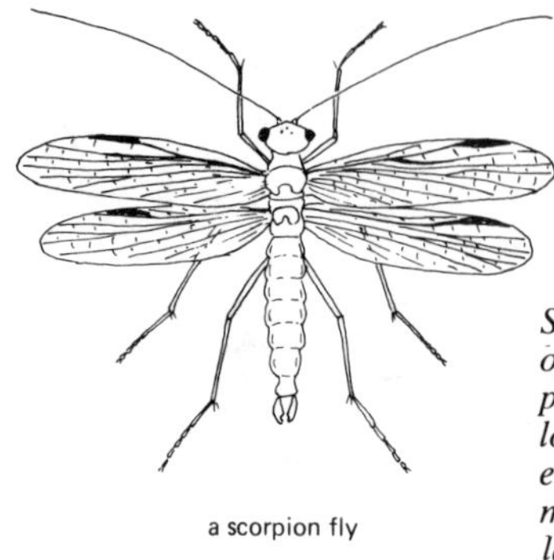

a scorpion fly

0.5 cm

Small to moderate sized insects with two pairs of large similar membranous wings showing a pattern of dark markings. Antennae rather long and filamentous. Biting mouthparts at the end of a snout. Short cerci present. Development holometabolous with caterpillar-like larvae.

The scorpion flies are slow flying insects usually found in moist situations or close to streams as the larvae of some species are aquatic. Some Mecoptera are plant feeding but others are predatory. They are of no economic significance.

ORDER 25. SIPHONAPTERA—fleas

(about 1,400 species world wide—about 60 in Great Britain)

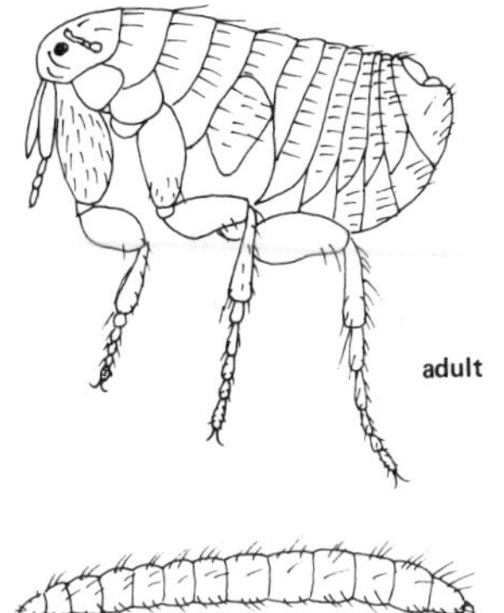

a flea

Small wingless hard bodied insects flattened from side to side. Adults ectoparasitic and blood feeding on warm blooded animals; larvae free living. Compound eyes absent, but one pair of ocelli usually present. Antennae short and thickened. Mouthparts modified for piercing and sucking. Legs adapted for jumping, with enlarged coxae. Development holometabolous with legless maggot-like larvae.

The adults of all species of fleas are ectoparasitic and blood sucking on higher animals whereas the larvae develop free from the host and feed on organic debris in the environs. Fleas are less host specific than lice so that fleas from the rats for instance, can transfer readily to humans, an important feature in relation to disease transmission. Since they can survive as adults for considerable periods of time away from a host, a house which has been empty for some time can immediately pose a problem with fleas when re-occupied.

Besides the irritant effect of their bites, fleas transmit several human diseases, bubonic plague in particular.

Main species of economic importance in Great Britain

Common name	*Scientific name*	*Principal hosts*	*Remarks*
cat flea	*Ctenocephalides felis*	cat, man	
dog flea	*C. canis*	dog, man	
European chicken flea	*Ceratophyllus gallinae*	poultry	
European rabbit flea	*Spilopsyllus cuniculi*	rabbit, cat	
human flea	*Pulex irritans*	man	
oriental rat flea	*Xenopsylla cheopis*	rat, man	vector of plague in the tropics
rat flea	*Nosopsyllus fasciatus*	rat	

ORDER 26. DIPTERA — flies

(about 150,000 species world wide — about 5,200 in Great Britain)*

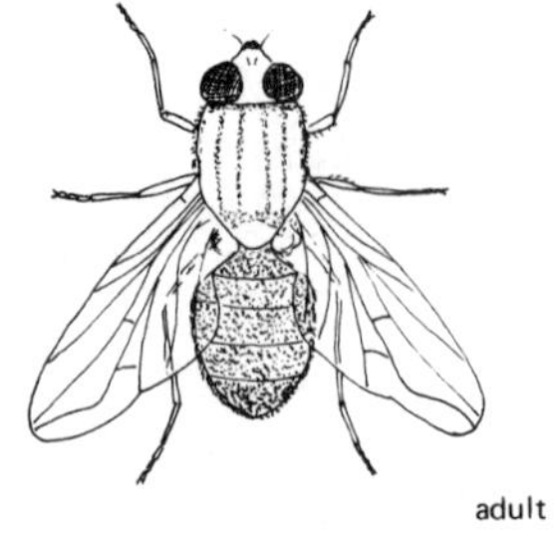

*Small to medium-sized insects with a single pair of membranous wings (forewings), the hindwings being modified into specialised balancing organs (***halteres***). Sponging or suctorial mouthparts sometimes adapted for piercing. Large mesothorax. Development holometabolous. Larvae legless, usually with reduced or retracted head. In higher forms the last larval skin is retained as an outer covering to the pupa forming a* **puparium**.

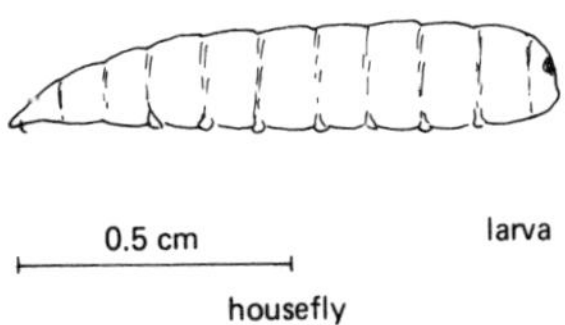

The Diptera are generally regarded as the most highly evolved order of insects. They are the only order to which the name flies should be applied. The adults are easily recognised as they possess only one pair of wings. The biology of Diptera is extremely diverse, particularly regarding the habitat and feeding behaviour of the larvae. Some are scavengers, others attack living plants, while others again are parasitic on other insects or on higher animals. Many adult flies are flower visitors, feeding on nectar or pollen; others feed on decaying organic matter whilst some are predatory on other insects or blood feeding on higher animals.

Because of this great variety of habit the Diptera are extremely important medically and in public health, as pests of livestock, and as pests of cultivated plants. Many of the most serious diseases of man in tropical countries, such as malaria, yellow fever, sleeping sickness and elephantiasis are transmitted by blood sucking Diptera.

*— See footnote at bottom of page 75

Main species of economic importance in Great Britain

Common name	*Scientific name*	*Principal hosts*	*Remarks*
Pests of plants			
apple leaf midge	*Dasineura mali*	apple	attacked leaves roll towards the midrib
bean seed fly	*Delia platura*	dwarf, runner beans, freesia	larvae tunnel in cotyledons and shoots of young plants
beet leaf miner (mangold fly)	*Pegomyia hyoscami*	sugar beet, mangold, spinach	larvae mine in leaves
blackcurrant leaf midge	*Dasineura tetensi*	blackcurrant	attacked leaves become twisted and folded
brassica pod midge	*Dasineura brassicae*	brassica seed pods	affects seed production as causes pods to shatter prematurely
cabbage root fly	*Delia brassicae*	cabbage, cauliflower and other brassicae	larvae attack roots
carnation fly	*Delia cardui*	carnation	larvae mine in leaves
carrot fly	*Psila rosae*	carrot, parsnip	larvae tunnel in tap root
celery fly	*Euleia heraclei*	celery, parsnip	larvae mine in leaves
cereal flies	*Opomyza* spp.	cereals, grasses	larvae feed on central shoot
chrysanthemum blotch miner	*Trypeta zoe*	chrysanthemum	larvae produce broad blotch-like mines
chrysanthemum gall midge	*Rhopalomyia chrysanthemi*	chrysanthemum	cone shaped galls produced on leaves and stems; now rare
chrysanthemum leaf miner	*Phytomyza syngenesiae* *P. horticola*	chrysanthemum	larvae mine in leaves
chrysanthemum stool miner	*Psila nigricornis*	chrysanthemum, lettuce	larvae tunnel in base of plant
clover seed midge	*Dasineura leguminicola*	red clover	larvae feed on developing seeds
foxtail midges (cocksfoot midges)	*Contarinia merceri* *Dasineura alopecuri*	meadow foxtail	reduce seed production
frit fly	*Oscinella frit*	oats, wheat, maize	larvae damage seedlings
gout fly	*Chlorops pumilionis*	barley, wheat, rye	causes characteristic swelling of infested shoots
large narcissus fly	*Merodon equestris*	narcissus	larvae destroy centre of bulb
leatherjackets	*Tipula paludosa* and other spp. (larvae)	pasture, turf, cereals	larvae feed on roots and underground stems

Common name	*Scientific name*	*Principal hosts*	*Remarks*
Pests of plants — *(continued)*			
lucerne flower midge	*Contarinia medicaginis*	lucerne	larvae attack flower buds and reduce seed yield
mushroom cecid	*Heteropeza pygmaea*	mushrooms	larvae tunnel into cap and stalk
mushroom sciarid flies	*Bradysia brunnipes* *B. paupera* *Lycoriella auripila* *L. solani* and other spp.	mushrooms, cucumber, carnations	larvae tunnel into mushrooms, attack roots and stems of other plants
onion fly	*Delia antiqua*	onion, leek, shallot	larvae tunnel into base of plant
pea leaf miners	*Liriomyza congesta* *L. pisivora*	pea	larvae mine in leaves
pea midge	*Contarinia pisi*	pea	leading shoots and flower buds deformed
pear leaf midge	*Dasineuri pyri*	pear	attacked leaves roll towards the midrib
pear midge	*Contarinia pyrivora*	pear	larvae feed inside young fruits
red bud borer	*Reseliella oculiperda*	apple, pear, plum, peach, rose	larvae attack budded plants in the nursery
raspberry cane midge	*R. theobaldi*	raspberry	larvae feed under rind of shoots
saddle gall midge	*Haplodiplosis marginata*	wheat, barley, rye	saddle-shaped galls develop on stems
small narcissus flies	*Eumerus strigatus* *E. tuberculatus*	narcissus and many other bulbs	larvae feed inside bulbs; usually secondary
spinach stem fly	*Delia echinata*	spinach	larvae mine in leaves
swede midge	*Contarinia nasturtii*	cruciferae	larvae attack growing points causing extensive distortion
timothy flies	*Nanna* spp.	timothy	larvae feed on developing seed
tomato leaf miner	*Liriomyza bryoniae*	tomato and other solanaceous plants	larvae mine in leaves
turnip root fly	*Delia floralis*	turnip, swede	larvae tunnel into heart of plant
violet leaf midge	*Dasineura affinis*	violet	larvae severely distort foliage
wheat blossom midges	*Contarinia tritici* *Sitodiplosis mosellana*	wheat, rye, barley	larvae attack seed heads
wheat bulb fly	*Delia coarctata*	wheat and occasionally barley	serious pest of winter wheat
white clover seed midge	*Dasineura gentneri*	white and red clover	larvae feed on developing seeds

Common name	*Scientific name*	*Principal hosts/prey*	*Remarks*
Predators of plant pests			
hover flies	*Melanostoma* spp. *Platycheirus manicatus* *Scaeva pyrastri* *Syrphus* spp.	aphids	only larvae are predatory
Parasites (of other insects)			
Tachinid flies	spp. of *Carcelia* *Compsilura* *Dexia Tachina* and other genera	larvae of Lepidoptera	only larvae are parasitic
Pests of livestock			
autumn fly	*Musca autumnalis*	cattle, horse	serious nuisance pest on faces of livestock; non biting; transmits *Thelazia* eye-worms
bot flies	*Gasterophilus* spp.	horse	larvae complete development in stomach or nasal cavities of host
blinding breeze fly	*Chrysops caecutiens*	cattle, sheep, horse	female is blood feeding, especially round the eye
blue bottles	*Calliphora vicina* *C. vomitoria*	sheep	eggs laid in soiled wool, larvae may attack flesh; also breed in meat and carrion
cattle biting fly	*Haematobosca stimulans*	cattle	common, blood feeding
common clegs	*Haematopota crassicornis* *H. pluvialis*	cattle, horse	common; may be involved in disease transmission
deer fly	*Lipotena cervi*	deer	
deer nostril fly	*Cephenemyia auribarbis*	deer	only in Scotland
deer warble fly	*Hypoderma diana*	deer	common in Scotland
forest fly	*Hippobosca equina*	horse	"New Forest fly"
horn fly	*Haematobia irritans*	cattle	blood feeding
horse flies	*Tabanus* spp. *Hybomitra* spp.	horse, cattle, sheep, pig, man	blood feeding females inflict painful bite
ox warble flies	*Hypoderma bovis* *H. lineatum*	cattle	larvae feed under skin of back and emerge through it to damage hide; *notifiable* pest
sheep head fly	*Hydrotaea irritans*	cattle, sheep, horse, deer	non-biting blood feeder; carries diseases and can cause large lesions especially in sheep

Common name	Scientific name	Principal hosts	Remarks
sheep ked	*Melophagus ovinus*	sheep	adult is blood feeding, wingless and flattened
sheep maggot fly	*Lucilia sericata*	sheep	major cause of primary "fly strike"
sheep nostril fly	*Oestrus ovis*	sheep	larvae occur in nostrils
stable fly	*Stomoxys calcitrans*	cattle, horse, sheep, pig, man	adults blood feeding
Pests of public health importance			
biting midges	*Culicoides* spp.		inflict painful bite but do not transmit disease in UK
black flies	Simuliidae		inflict painful bite but do not transmit disease in UK; swarms can kill cattle in Europe through blood loss
cheese skipper	*Piophila casei*	larvae infest cheese, bacon, ham	Larvae jump when disturbed
cluster and swarming flies	*Dasyphora* spp. *Pollenia* spp. *Thaumatomyia* spp.		adults may collect in buildings in large numbers in spring and autumn
flesh flies	*Sarcophaga* spp.		breed in material of animal origin; together with blue bottles cause fly blown meat
house fly	*Musca domestica*		main fly pest in houses; breeds in decaying organic matter of all kinds
latrine fly	*Fannia scalaris*		breeds in privies
lesser house fly	*F. canicularis*		may breed in large numbers in poultry houses
mosquitoes	*Culex* spp. *Aedes* spp. and other genera		may cause irritant bites but generally do not transmit disease organisms in Great Britain
moth flies	Psychodidae		may swarm in large numbers but harmless
non-biting midges	Chironomidae		nuisance only
small fruit flies	*Drosophila* spp.		breed in over ripe fruit and also in milk curds
window gnats (sewage filter-bed flies)	Anisopidae		may swarm in large numbers but harmless

ORDER 27. LEPIDOPTERA — butterflies, moths

(about 112,000 species world wide—about 2,300 in Great Britain)*

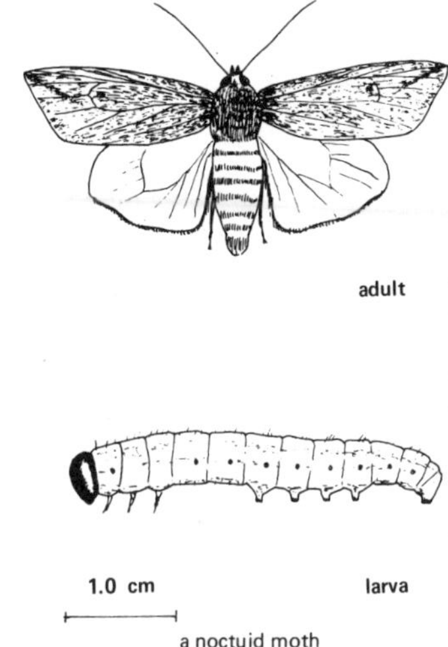

a noctuid moth

Small to large insects with two pairs of large membranous wings covered in minute flattened scales. Body and legs also clothed in scales and hairs. Mouthparts consist of a long tubular proboscis formed from the maxillae. Eyes well developed. Antennae clubbed, tapering or feathery. Development holometabolous. Larvae are soft-bodied caterpillars with a hard head capsule and well developed biting mouthparts. Three pairs of true jointed legs with single claws present at fore-end. Additional false legs present on the abdomen with groups of small spines in place of claws. Pupae with limbs smoothly enclosed; usually in a silken cocoon or earthern cell.

The Lepidoptera, which comprise the well-known butterflies and moths, are a large order of insects that occur throughout the world in both warm and temperate climates. In tropical areas they occur in great profusion of species, size and colour. The order is almost exclusively plant feeding in the larval stage and includes some of the most serious pests of plants in the world. Almost every kind of cultivated plant has one or more lepidopterous pests that attack it. Adult butterflies and moths feed mostly on nectar from flowers, though some do not feed at all, and are thus almost without exception quite harmless. However, a few species of tropical moths in the adult stage can injure ripe fruits by piercing the skin. This habit does not occur with any of the temperate species.

Larvae of butterflies and moths (caterpillars), possess biting mouthparts and injure plants by consuming plant tissue of one sort or another. Many species feed openly on leaves but others bore into stems or fruits, or tunnel within leaves, while a few are root feeding. Some Lepidoptera, along with the Coleoptera (beetles), are important pests of stored foods. Others damage woollen goods, including clothing and carpets.

*— See footnote at bottom of page 75

Main species of economic importance in Great Britain

Common name	*Scientific name*	*Principal hosts*	*Remarks*
Pests of plants			
angle shades moth	*Phlogophora meticulosa*	tomato, carnation, chrysanthemum, cruciferae	mainly under glass
azalea leaf miner	*Caloptilia azaleella*	azalea	larvae mine in and severely damage foliage
bramble shoot moth	*Epiblema uddmanniana*	blackberry, loganberry	larvae web leaves together
brown-tail moth	*Euproctis chrysorrhea*	many trees and shrubs	larvae live in colonies and form webbing; shed irritant hairs
bud moth	*Spilonota ocellana*	apple	larvae feed on opening buds and blossoms
cabbage moth	*Mamestra brassicae*	cabbage and other brassicas	larvae skeletonise leaves
carnation tortrix	*Cacoeciomorpha pronubana*	carnation	larvae attack foliage and flower buds
cherry fruit tortrix	*Argyresthia curvella*	cherry	larvae feed on flower buds, flowers and young fruits
codling moth	*Cydia pomonella*	apple, pear, quince, walnut	larvae tunnel in fruit
currant clearwing	*Syanthedon tipuliformis*	currant, gooseberry	larvae tunnel in stems
cutworms	*Argotis segetum* *A. exclamationis* *Noctua pronuba* *Euxoa nigricans*	many plants, important in potatoes, carrots, lettuce	larvae eat through stem at ground level; epidemics infrequent
diamond-back moth	*Plutella xylostella*	cabbage and other brassicas	larvae eat "windows" in leaves
fruit tree tortrix	*Archips podana*	apple	larvae may eat deep irregular holes in maturing fruit but mostly foliage feeding
fruitlet mining tortrix	*Pammene rhediella*	apple, plum	larvae eat small holes in fruitlets
grass moths	Crambinae	pasture grasses, cereals	plants eaten at or below soil level
lackey moth	*Malacosoma neustria*	many trees and shrubs	larvae live in colonies and form webbing on foliage
large white butterfly	*Pieris brassicae*	cabbage and other brassicas	larvae skeletonise leaves
leek moth	*Acrolepiopsis assectella*	leek, onion, shallot	larvae bore through folded leaves

Common name	*Scientific name*	*Principal hosts*	*Remarks*
Pests of plants — *(continued)*			
leopard moth	*Zeuzera pyrina*	apple and other woody plants	larvae bore in shoots
lilac leaf miner	*Caloptilia syringella*	lilac, privet	larvae mine in leaves
magpie moth	*Abraxas grossulariata*	gooseberry, currants	larvae feed on foliage
March moth	*Alsophila aescularia*	apple and other fruits	larvae perforate foliage and may scar fruits
Mediterranean climbing cutworm	*Spodoptera littoralis*	chrysanthemum	larvae damage foliage, flowers and buds under glass; not indigenous, occasionally imported
mottled umber	*Erannis defoliaria*	apple and other fruits	larvae perforate foliage and may scar fruits
parsnip moth	*Depressaria pastinacella*	parsnip, carrot, celery	can seriously depress seed production
pea moth	*Cydia nigricana*	pea	serious pest of peas, larvae feed on peas within pod
pine looper	*Bupalus piniarius*	pine	larvae feed on needles
pine shoot moth	*Rhyacionia buoliana*	pine	
plum fruit tortrix	*Cydia funebrana*	plum	larvae tunnel into fruits
plum tortrix	*Hedya pruniana*	plum	larvae feed on foliage and tunnel into shoots
raspberry moth	*Lampronia rubiella*	raspberry, longanberry, blackberry	larvae tunnel into buds
rosy rustic moth	*Hydraecia micacea*	potato and many other plants	larvae feed on stems at ground level
silver Y moth	*Autographa gamma*	beans, chrysan--themum, sugar beet	especially under glass
small ermine moths	*Yponomeuta* spp.	many trees and shrubs	larvae live in colonies and form webbing on foliage
small white butterfly	*Pieris rapae*	cabbage and other brassicas	larvae skeletonise foliage
strawberry tortrix	*Acleris comariana*	strawberry	larvae feed on foliage and flowers
summer fruit tortrix	*Adoxophyes orana*	apple, pear	larvae cause large shallow depressions on fruit
swift moths	*Hepialus humuli* *H. lupulinus*	many plants	larvae live in soil and feed on roots and tubers

Common name	*Scientific name*	*Principal hosts*	*Remarks*
Pests of plants — *(continued)*			
tomato moth	*Laconobia oleracea*	tomato, carnation, chrysanthemum, cruciferae	mainly under glass
vapourer moth	*Orgyia antiqua*	many trees and shrubs	larvae feed on foliage
winter moth	*Operophtera brumata*	apple	larvae perforate foliage and may scar fruit
Pests of stored products			
brown house moth	*Hofmannophila pseudospretella*	grain products and woollen goods	larvae require high humidity
clothes moth, brown-dotted	*Niditinea fuscipunctella*	woollen clothing, carpets	common in poultry houses
clothes moth, case bearing	*Tinea pellionella*	woollen clothing, carpets	
clothes moth, common	*Tineola bisselliella*	woollen clothing, carpets	
clothes moth, large pale	*Tinea pallescentella*	woollen clothing, carpets	
clothes moth, white tip	*Trichophaga tapetzella*	woollen clothing, carpets	
corn moth	*Nemapogon granella*	grain and grain products	
Indian meal moth	*Plodia interpunctella*	dried fruits, nuts, cereals, oilseeds and products	can overwinter but generally requires heated premises
meal moth	*Pyralis farinalis*	milled grain and flour products	usually a minor pest; in damp situations
Mediterranean flour moth	*Ephestia kuehniella*	flour and cereal products	principal moth pest of flour mills
tropical warehouse moth	*E. cautella*	dried fruits and vegetables, nuts, oilseeds and products, cocoa	requires heated premises
warehouse moth	*E. elutella*	wide range of stored products including grain, chocolate, dried fruits and vegetables, nuts and tobacco	
white shouldered house moth	*Endrosis sarcitrella*	woollen goods and many cereal products	especially in damp situations

ORDER 28. TRICHOPTERA — caddis flies

(about 5,000 species world wide — 189 in Great Britain)

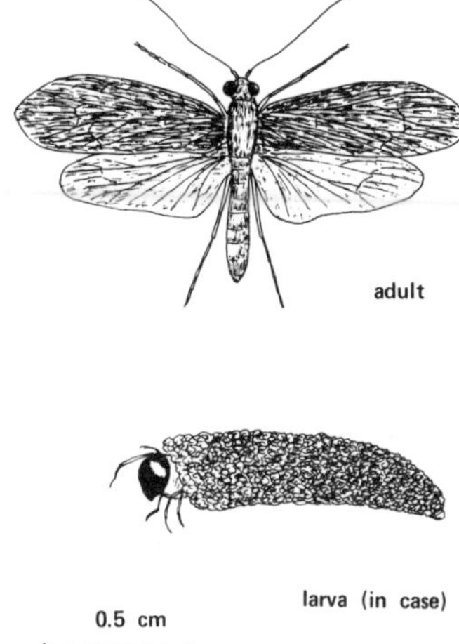

a caddis fly

Small to moderate-sized moth-like insects with two pairs of large membranous wings densely covered with fine hairs. Mouthparts with mandibles greatly reduced or absent. Antennae long and gently tapering. Development holometabolous. Larvae aquatic in fresh, mostly flowing water; caterpillar-like and usually within a case constructed of small stones, sand, leaves or twigs, External abdominal gills usually present.

Caddis flies are probably better known as larvae than as adults. They occur in huge numbers in streams and rivers and form an important item in the diet of freshwater fish. Adult caddis flies are often mistaken for moths, but can be readily distinguished by the presence of hairs on the wings instead of scales, and the form of the mouthparts which lack the coiled proboscis typical of Lepidoptera. They are feeble nocturnal fliers and do not stray far from fresh water. They may be attracted to light in large numbers. A few species have been recorded as being destructive to water lilies and other aquatic plants but generally the order is of no economic significance other than as a food source for trout and other freshwater fish.

ORDER 29. HYMENOPTERA — sawflies, ants, bees, wasps

(about 110,000 species world wide — about 6,000 in Great Britain)*

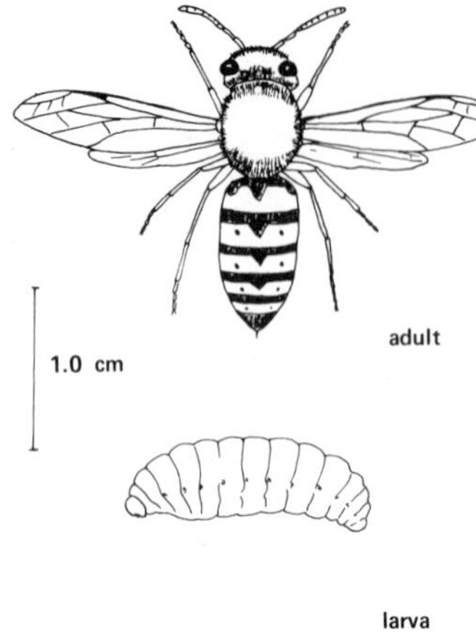

German wasp

Minute to moderate-sized insects with two pairs of membranous wings. The hindwings always smaller than the forewings to which they are fastened by a series of hooks. Veins greatly reduced in the smaller forms. Mouthparts primarily of the biting pattern but may be modified for lapping. A marked constriction (waist) apparently between the thorax and abdomen (except in sawflies). Ovipositor always present and modified for piercing or stinging. Development holometabolous. Larvae soft-bodied, and legless, except the larvae of sawflies which have a hard head capsule and possess both true and false legs.

The Hymenoptera are more important for their beneficial activities than harmful ones. These include not only pollination of plants and production of honey (by bees in particular) but also parasitism of other insects which is a very common habit amongst the Hymenoptera. Many of the most valuable natural enemies of pest insects come from this group.

Some Hymenoptera are however plant feeding (the sawflies are entirely so) so that the order is not without its plant pests. In addition, many ants and some wasps are nuisance pests around the home and elsewhere.

Many species of Hymenoptera are social in habit in that they live in large complex colonies with specialisation of reproductive function and of other activities. In behaviour they are amongst the most highly evolved species. All of the parasitic species are however solitary in habit.

*— See footnote at bottom of page 75

Main species of economic importance in Great Britain

Common name	*Scientific name*	*Principal hosts*	*Remarks*
Pests of plants			
Sawflies			
antler sawflies	*Cladius pectinicornis* *C. difformis*	rose	larvae eat irregular holes in leaves
apple sawfly	*Hoplocampa testudinea*	apple	larvae tunnel in fruit
banded rose sawfly	*Allantus cinctus*	rose	larvae eat holes in leaves
blackcurrant sawfly	*Nematus olfaciens*	gooseberry, currants	larvae feed on foliage
common gooseberry sawfly	*Nematus ribesii*	gooseberry, currants	larvae feed on foliage
large rose sawfly	*Arge achropus*	rose	larvae eat holes in leaves
leaf-rolling rose sawfly	*Blennocampa pusilla*	rose	attacked leaves become tightly rolled
pale spotted gooseberry sawfly	*Nematus leucotrochus*	gooseberry, currant	larvae feed on foliage
pear sawfly	*Hoplocampa brevis*	pear	larvae tunnel in fruit
pear slug sawfly	*Caliroa cerasi*	pear, cherry, apple, hawthorn	larvae consume one surface of leaf only
plum sawfly	*Hoplocampa flava*	plum	larvae tunnel in fruit
rose slug sawfly	*Endelomyia aethiops*	rose	larvae consume one surface of leaf only
turnip sawfly	*Athalia rosae*	turnip, swede	larvae skeletonise leaves
Other plant pests			
Douglas fir seed wasp	*Megastigmus spermatrophus*	Douglas fir	reduces seed yield
leaf cutter bees	*Megachile* spp.	rose	adult females cut circular portions from leaves
Nuisance pests			
Ants			
common black ant	*Lasius niger*		usually nests outside in lawns, flower beds etc.; invades houses and is attracted to sweet foods

Main species of economic importance in Great Britain

Common name	*Scientific name*	*Principal hosts*	*Remarks*
Nuisance pests — *(continued)*			
Pharaoh's ant	*Monomorium pharaonis*		requires artificial heat for breeding; nests in wall cavities, around heating pipes etc.; especially in institutions and hospitals
Wasps			
common wasp	*Paravespula vulgaris*		principally nuisance pests but may attack ripe fruits
German wasp	*P. germanica*		
Pollinators			
bumble-bees	*Bombus* spp.		
honey bee	*Apis mellifera*		
Parasites* (of other insects)			
aphid parasitic wasps	*Aphelinus* spp. *Aphidius* spp.	aphids, scale insects	
cabbage aphid parasite	*Diaeretiella rapae*	cabbage aphid	
carrot fly parasite	*Chorebus gracilis*	carrot fly larva and pupa	
common apanteles	*Apanteles glomeratus*	white butterflies (larvae)	
diamond back moth parasite	*Diadegma fenestralis*	diamond back moth larva	
glasshouse whitefly parasite	*Encarsia formosa*	"scales" of whitefly	
leaf miner parasites	*Diglyphus* spp. *Dacnusa* spp.	chrysanthemum leaf miner	being reared for control of chrysanthemum leaf miner
red ichneumon wasps	*Ophion* spp.	Lepidopterous larvae	
white butterfly parasitic wasp	*Pteromalus puparum*	white butterfly pupa	
wood wasp parasite	*Rhyssa persuasoria*	*Sirex* wasp larva	
—	*Idiomorpha rapae*	cabbage root fly larva	
—	*Trichogramma* spp.	Lepidopterous eggs	

*Note: Only a very few common species of parasitic Hymenoptera are listed here. Many more exist which do not have common names but which may be important as natural enemies of pest species.

SELECTED REFERENCES

Chinery, M. 1979, *A field guide to the insects of Britain and Northern Europe*. Collins, London. 352 pp.

Kloet, G. S. Hincks, W. D. 1945. *A check list of British insects*. Kloet and Hincks, Stockport. 483 pp.

Ministry of Agriculture, Fisheries and Food. Agricultural Development and Advisory Service. Advisory leaflets on individual pests or pest groups. Pest identification cards. Her Majesty's Stationery Office, London.

Richards, O. W., Davies, R. G. 1977. *Imm's general textbook of entomology*. 10th edition. Vol. 2. Classification and biology. Chapman and Hall, London. 933 pp.

Royal Entomological Society. Handbooks for the identification of British Insects. Royal Entomological Society, London.

Seymour, P. R. 1979. *Invertebrates of economic importance in Britain. Common and scientific names*. Ministry of Agriculture, Fisheries and Food. London. 132 pp.

Chapter 7

Mites, and Other Non-Insect Pests

Entomology in the strict sense is concerned only with insects but in practice the entomologist is often expected to deal with other invertebrate animals which may be troublesome as pests of plants or higher animals. This applies particularly to mites which like insects are arthropods, but also sometimes to non-arthropod pests such as slugs, snails and nematodes. This chapter deals briefly with these non-insect groups, with the exception of nematodes. The latter are now recognised as extremely important pests of plants in some situations and their study has developed into a distinct branch of applied science called nematology. Because of its specialised nature it is not considered in this book.

The classification, scientific names, common names and comments on the practical importance of these various non-insect groups are set out in Table 7. The reader is also referred to Fig 1 for their overall position in the animal kingdom.

Mites

(a) *General features*

Mites are extremely varied in lifestyle. Many species are scavengers feeding on dead and decaying remains of plants and animals and may therefore play an important role in the early stages of decay and re-cycling processes. Others feed on living plants and these include some very important pests of cultivated plants. A few species of mites are predatory on other mites and some affect higher animals. Mites together with ticks constitute the order **Acari** and are closely related to the spiders (**Araneae**). These two groups together form the class **Arachnida**. All mites are small, most species being 1 mm or less in length, while some are smaller still. Ticks on the other hand, which are similar to mites in many ways, may be 1 cm or more in length when fully grown. All ticks are ectoparasitic and blood feeding on vertebrate animals at some stage in their life cycle.

Table 7. Non-insect invertebrates*: classification, common names and practical importance

Phylum	*Class*	*Order*	*Common name(s)*	*Practical importance*
Arthropoda	Arachnida	Acari	mites and ticks	Many species of mites are important pests of cultivated plants, others are scavengers and some are predatory on other mites; ticks are ectoparasites of higher animals.
		Araneae	spiders	Predatory on insects and other small animals.
	Myriapoda	Chilopoda	centipedes	Predatory on insects and other small animals.
		Diplopoda	millipedes	Minor plant pests; mostly feed on dead plant material.
		Symphyla	symphilids	Sometimes important pests of cultivated plants; root feeding.
	Crustacea	Isopoda	slaters or woodlice	Minor plant pests; mostly feed on dead plant material.
Mollusca	Gastropoda	Pulmonata	slugs and snails	Important pests of cultivated plants.

* **excluding nematodes**

Mites are arthropods so that there is considerable similarity to insects, the most obvious features being the possession of an external jointed skeleton and distinctly jointed legs. In common with all other arthropods, mites moult at intervals as they grow. However, there are important points of distinction from insects, as detailed below and in Table 8.

In all arachnids the body segments are grouped in a different way compared to insects. Instead of the head, thorax and abdomen of insects there are only two body regions — a forward **prosoma** composed of six segments and usually bearing four pairs of walking legs, and a hind **opisthosoma** of thirteen segments without legs. This separation into two body regions is quite distinct in spiders (Araneae) but in most mites and ticks (Acari) the two regions have become merged so that they present a single rounded appearance. Also there is very little external evidence of segmentation. The general appearance of a typical mite is shown in Fig 16.

Mites typically possess four pairs of walking legs which arise from the forepart of the body. However, some mites have only two pairs of legs (particularly the Eriophyid gall mites). In all other mites the first instar stage possesses three pairs of legs whereas later stages possess four pairs. Mites do not possess antennae but in some cases the first pair of legs may be sensory in function. Chemical sense receptors are also associated with the mouthparts. Compound eyes are never present but there may be several pairs of simple eyes similar to those of insects. However, many mites are blind.

Table 8. Main points of similarity and distinction between mites and insects

Feature	Mites	Insects
Points of similarity		
external skeleton	yes	yes
jointed legs	yes	yes
tracheal system	some only	yes
moulting takes place	yes	yes
Points of distinction		
body regions	one (slightly divided into two in some groups)	three (head, thorax, abdomen)
antennae	absent	present
legs	four pairs (in most cases)	three pairs

(b) *Mouthparts*

The mouthparts of mites are of either a chewing or piercing form but do not correspond in any way to those of insects. They are

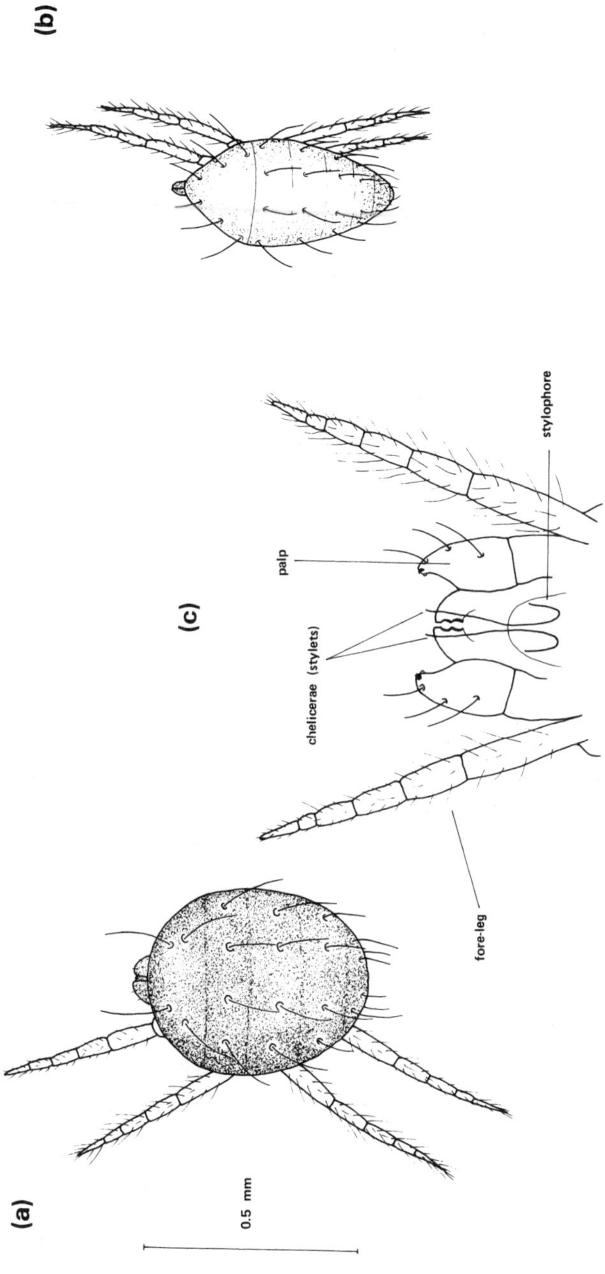

Fig. 16. A typical plant feeding mite (fruit tree red spider mite); (a) adult female, (b) adult male, (c) mouthpart structure.

formed from two pairs of appendages in the head region which are sometimes borne on a projection referred to as a false head (**capitulum**). In chewing forms a pair of pincer-like **chelicerae** is present. In mites with piercing mouthparts stylet-like structures may be developed as part of the feeding apparatus (see Fig 16 (c)). Plant feeding mites are essentially liquid feeders on plant-juices and have mouthparts adapted for this purpose.

(c) *Respiratory system*

A simple tracheal system of respiration is present in many mites, which are one of the few groups of organisms besides insects to respire in this way. The number of spiracles present (there may be only a single pair) and their arrangement are features used in the classification of mites. Many very small mites do not possess a tracheal system and respire through the general body surface.

(d) *Life cycles*

For the most part reproduction is sexual, sometimes with a distinct difference in appearance between male and female (Fig 16). However in some species parthenogenesis may be the normal or occasional mode of reproduction. Eggs are normally laid but some species are viviparous. Eggs are simple in form and are usually smoothly rounded or oval. Plant feeding mites normally attach their eggs to the surface of host plants.

Development from egg to adult is simple with little metamorphosis, a miniature version of the adult hatching out from the egg. However, this first stage possesses only three pairs of legs compared to the adult's normal four pairs. (The life cycle of fruit tree red spider mite is shown diagrammatically in Fig 17.) Terms applied to the juvenile stages of mites are the same as those used for immature insects. Thus the first stage is referred to as a "larva" and the following stages prior to adulthood as "nymphs". In the author's opinion this terminology is unfortunate as it leads to confusion with the immature stages of insects to which the terms correctly apply. Most mites undergo three or four moults before they are adult. Just prior to each moult they enter an immobile resting stage in which the legs are tucked up close to the body. The life cycle of most mites is short, in fruit tree red spider mite taking only about ten days under summer conditions.

Some mites can develop a nymphal stage specialised for dispersal and called a **hypopus**. This stage is non-feeding and possesses suckers or claws for attaching itself to other animals. In this way it can be carried considerable distances.

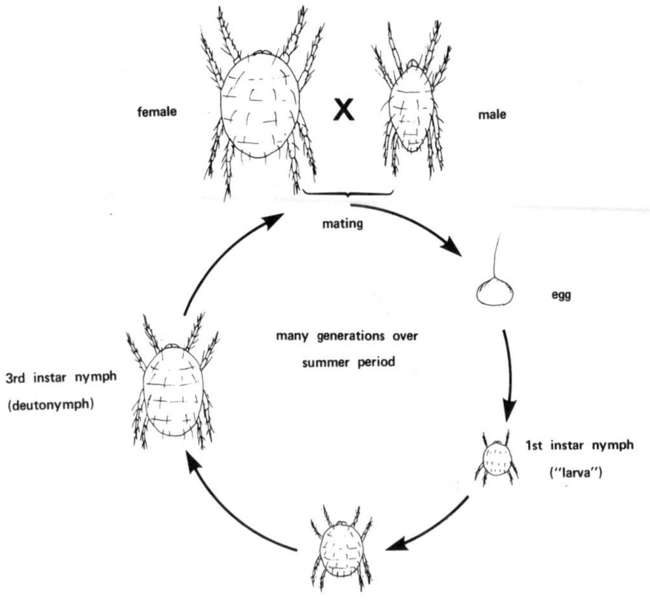

Fig. 17. Life cycle of fruit tree red spider mite.

(e) *Main types of mites*

For most purposes detailed knowledge of classification of mites is not necessary though it is important to be able to recognise the main groups of plant feeding mites and their damage symptoms. Where integrated pest management of plant feeding mites (see Chapter 12) is being attempted it is also essential to distinguish between predatory and plant feeding species.

There are four main groups of mites amongst which plant damaging species occur. These are listed in Table 9 together with their main features and comments on practical importance. The general appearance of representative species from important groups of mites and a tick are shown in Fig 18. The more important species of mites and ticks recorded from Great Britain are listed in Table 10.

(f) *Biology of a typical phytophagous mite*

Fruit tree red spider mite (*Panonychus ulmi*) may be taken as an example of typical plant feeding mite. It is an important pest of

Table 9. The main groups of mites important as plant pests, predators or scavengers

Family	*Common name(s)*	*Main features*	*Length when fully grown*	*Practical importance*
Plant pests				
Tetranychidae	tetranychid mites (spider mites)	Comparatively large mites, generally globular in form with prominent body bristles; four pairs of similar legs; prominent capitulum.	up to 1.0 mm	Serious plant pests; feeding results in flecking and bleaching of foliage.
Tarsonemidae	tarsonemid mites	Very small mites of rounded shape with four pairs of legs but the hind pair very slender and bristle-like; body shiny and without obvious bristles; capitulum greatly reduced.	up to 0.25 mm	Occasionally serious plant pests; feeding causes distortion and stunting.
Eriophyidae	eriophyid mites (erineum mites, bud mites, rust mites,	Minute, elongate, colourless, worm-like mites with two pairs of very short legs anteriorly placed; prominent capitulum.	up to 0.20 mm	Serious pests of plants; feeding causes distortion, stunting or gall formation; many similar species.

Acaridae	acarid mites (tyroglyphid) mites)	Comparatively large globular mites with body slightly divided by a transverse groove; four pairs of short, stubby legs; movement sluggish	up to 0.8 mm	Minor plant pests, usually associated with rot fungi; many species are pests of stored foods.
Predators				
Phytoseiidae	phytoseiid mites (typhlodromid mites, amblyseiid mites)	Moderate sized, very active mites with four pairs of promintent legs; body often pear-shaped.	up to 0.5 mm	Important predators of spider mites and may provide effective control if managed; feeding preferences often specific.
Scavengers				
Oribatei	oribatid mites (beetle mites, soil mites)	Moderate to large-sized mites; heavily built and often dark brown in colour; "beetle-like" in appearance; few or no body bristles.	up to 0.8 mm	Feed on fungi and dead and decaying plant material; very common in leaf mould and surface layers of the soil.

Table 10. The more important species of mites and ticks in Great Britain

Common name	*Scientific name*	*Principal hosts*	*Remarks*
Plant pests			
apple rust mite	*Aculus schlechtendali*	apple	may russet surface of leaves and fruit
blackberry mite	*Acalitus essigi*	blackberry	probable cause of "redberry" condition
blackcurrant gall mite	*Cecidophyopsis ribis*	blackcurrant	cause of "big bud" condition, transmits reversion virus
broad mite	*Polyphagotarsonemus latus*	ornamental plants under glass	attacked leaves become deformed and curled at edges
bryobia mite	*Bryobia* spp.	apple, pear, gooseberry	bronze leaves
bulb mites	*Rhizoglyphus callae* *R. robini*	narcissus and other bulbs	occur in association with rot fungi
bulb scale mite	*Steneotarsonemus laticeps*	narcissus	serious pest of forced narcissus
carmine spider mite	*Tetranychus cinnabarinus*	carnation, arum and occasionally tomatoes	adult females dark red in colour
chrysanthemum leaf rust mite	*Epitrimerus alinae*	chrysanthemum	russets stem particularly just below flower head
conifer spinning mite	*Oligonychus ununguis*	conifers	attacked foliage yellows and withers
fern mite	*Hemitarsonemus tepidariorum*	ferns	fronds become distorted and stunted
fruit tree red spider mite	*Panonychus ulmi*	apple	attacked foliage becomes bronzed
grainstack mite (French fly)	*Tyrophagus longior*	cucumber	occurs in straw, hay and stored grain; may damage leaves and shoots of young plants
mould mite	*Tyrophagus putrescentiae*	mushrooms (among cultivated plants), but more common on stored foods	causes small irregular pits in caps and stalks; also attacks a wide range of stored foods

Table 10 — *(continued)*

Common name	*Scientific name*	*Principal hosts*	*Remarks*
Plant pests — *(continued)*			
mushroom mite	*Tarsonemus myceliophagus*	mushrooms	causes reddish brown discolouration at base of stipe
pear leaf blister mite	*Eriophyes pyri*	pear	causes small blisters on leaves; rare in commercial orchards
strawberry mite	*Tarsonemus pallidus*	strawberry, cyclamen	attacked leaves become deformed and curled at edges
tomato russet mite	*Aculops lycopersici*	tomato	may russet surface of fruit
two-spotted red spider mite	*Tetranychus urticae*	cucumber, tomato, strawberry, hops and a wide range of ornamentals	attacked leaves become bleached
vine leaf blister mite	*Colomerus vitis*	vines	causes erinea (blisters) on leaves
Predators of plant pests			
fruit tree red spider mite predator	*Typhlodromus pyri*	fruit tree red spider mite	can provide effective control
two-spotted red spider mite predator	*Phytoseiulus persimilis*	two-spotted red spider mite	being managed for control of spider mites under glass; originated from Chile
Parasites of man and animals			
cat head mange mite	*Notodedres cati*	cat	cause of mange in cats
chicken mite (poultry red mite)	*Dermanyssus gallinae*	poultry	common on poultry and wild birds
coastal red tick	*Haemaphysalis punctata*		
depluming mite	*Knemidokoptes gallinae*	poultry	infestation causes large bare patches

Table 10 — *(continued)*

Common name	*Scientific name*	*Principal hosts*	*Remarks*
Parasites of man and animals — *(continued)*			
dog fur mite	*Cheyletiella yasguri*	dog	
dog tick	*Ixodes canisuga*	dog	
ear mange mite	*Otodectes cyanotis*	cat, dog	lives in ear
follicle mites	*Demodex* spp.	cattle, sheep, dog, cat, man	various species occur
fowl tick	*Argas persicus*	poultry	
grain itch mite	*Pyemotes ventricosus*	insects	a predator of many insects but may cause allergic skin reaction in man
harvest mite	*Neotrombicula autumnalis*	man	causes skin irritation
itch mite	*Sarcoptes scabiei*	man, horse, sheep, cattle	cause of scabies in man and of *notifiable* mange in horses and sheep
psoroptic mange mite	*Psoroptes communis* (especially var. *ovis*)	sheep, cattle	cause of mange, especially "sheep scab", a *notifiable* "disease" of sheep
scaly-leg mite	*Knemidokoptes mutans*	poultry	causes lesions on legs
sheep chorioptic mange mite	*Chorioptes bovis*	sheep	cause of foot and body mange in sheep; may cause scrotal mange in rams
sheep tick	*Ixodes ricinus*	sheep, cattle, dog, man	important in disease transmission
Domestic and stored products pests			
cheese mite	*Tyrolichus casei*	old grain residues	rarely found on cheese despite the common name
cosmopolitan food mite	*Lepidoglyphus destructor*	wide range of stored foods	can cause dermatitis (grocer's itch)
dried fruit mite	*Carpoglyphus lactis*	dried fruits	
flour mite	*Acarus siro*	cereals and cereal products	
house mite	*Glycyphagus domesticus*	flour, flour products and mould growth	can cause dermatitis (grocer's itch)
house dust mites	*Dermatophagoides* spp.	minute fragments of organic matter in bedding etc.	cause of respiratory allergy in some people

deciduous fruit trees in many parts of the world. During the winter months eggs only (winter eggs) are present on the woody parts of the tree. When the leaves develop in the spring the eggs hatch and the tiny first stage mites move onto the foliage and commence feeding. There are three moults before adulthood with a short resting stage prior to each moult. The time from egg to adult is 10-14 days depending on conditions. Adult female mites are almost globular, deep brick red in colour with prominent light spines and about 0.5 mm in length. Males are rather smaller, pointed at the hind end and lighter in colour. After mating the females lay eggs which are attached to the leaf surface. These summer eggs are lighter in colour than the winter eggs and hatch after a few days. Development is rapid and five or more overlapping generations may occur over the summer. Prior to leaf fall in autumn females move to the woody parts of the tree and deposit winter eggs. No active stages live through the winter.

The mites feed by piercing the leaves and sucking out the cell contents of mainly the parenchyma tissue. This produces minute white flecks on the leaf which eventually coalesce resulting in a bleached or bronzed appearance. Heavily attacked leaves shrivel and drop prematurely. The effect on the tree is reduction of vigour and fruit size. Bud formation for the following season may also be affected.

(g) *Plant injury caused by mites*

Because of their small size the presence of mites may go unnoticed in the early stages of infestation. It is therefore extremely important to be able to recognise plant injury caused by mites at an early stage so that appropriate measures can be taken. Even with the larger mite species symptoms of attack are often apparent before the mites themselves are seen and in the case of very small mites, such as tarsonemids and eriophyids, this is invariably the case.

The nature of plant injury depends to a great extent on the type of mite involved and the species and part of the plant attacked. All spider mites (Tetranychidae) feed as described above for fruit tree red spider mite resulting in minute flecks appearing on leaves or other plant parts. Most feeding takes place from the undersides of leaves but injury is often seen more readily on the upper surface. Flowers may be attacked as well as foliage. There is usually little or no distortion of the plant parts attacked. Some species, in particular the two-spotted mite (*Tetranychus urticae*), produce considerable amounts of fine silken webbing over the plant surface, but other species of tetranychids, including fruit tree red spider mite, produce little or no webbing.

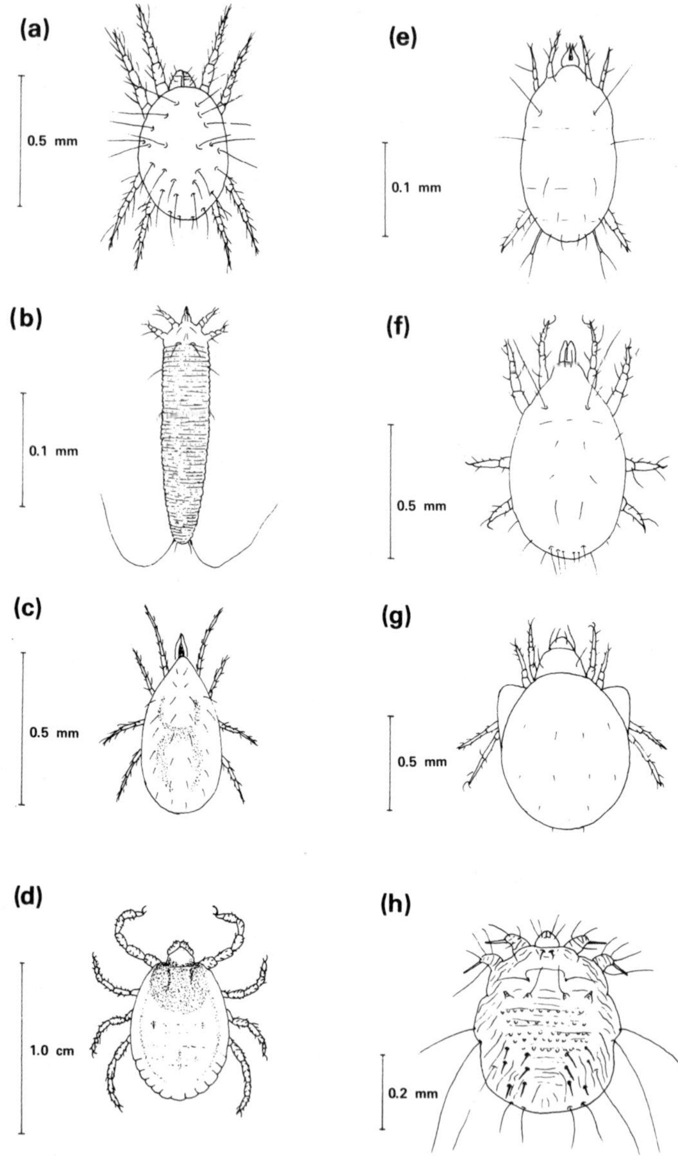

Fig. 18. Major types of mites and a tick. (All dorsal view. Note differences in scale);

(a) tetranychid mite,
(b) eriophyid mite,
(c) phytoseiid mite,
(d) tick,
(e) tarsonemid mite,
(f) acarid mite,
(g) oribatid mite,
(h) scabies mite.

In contrast, feeding by tarsonemid and eriophyid mites does not generally produce obvious bleaching of attacked areas. Instead considerable distortion, puckering or stunting of leaves or other plant parts is a common symptom. With some species of eriophyid mites distinct galls are produced. Mites will generally be found within the affected areas but high power microscope examination is required because of their minute size. An added difficulty is that most species of tarsonemid and eriophyid mites are colourless or semi-transparent.

Tyroglyphid mites are most likely to be met attacking bulbs. Affected bulbs grow poorly and are soft and spongy to the touch. Internally areas of rotting tissue are invariably found associated with the presence of the mites and it is usually uncertain which are the primary invaders. The mites tend to occur in clusters. They are readily visible to the naked eye as glistening white globules showing sluggish movement.

Millipedes and centipedes

Millipedes and centipedes belong in the Myriapod class of arthropods (see Table 7 and Fig 1). The name means many feet and both groups possess many body segments in the adult form each bearing legs, though early stages possess only a few segments. The general body form in both cases is a distinct rounded head and a long sinuous body composed of numerous similar segments. Millipedes and centipedes are however quite distinct and it is important to be able to distinguish between them as millipedes feed on plant material and may be troublesome pests, whereas centipedes are entirely predatory on insects and other small animals and hence beneficial.

The head of millipedes is globular and the body round in cross section. There are two pairs of legs to each body segment (except for the three segments immediately behind the head) and movement in life is sluggish (see Fig 19(a)). In contrast, the head and body of centipedes are rather flattened and each body segment bears a single pair of legs rather than two pairs (Fig 19(b)). Movement is quite rapid, often with a marked sinuous action.

Although millipedes feed entirely on plant material this is mostly in the form of dead and decaying plant remains. Under certain conditions however they will attack growing plants, especially soft succulent parts. They are favoured by the presence of large amounts of organic manure in proximity to the plants and by high moisture levels, but the precise nature of conditions which induce them to attack living plants are not well understood. Rot organisms are often associated with plant tissue injured by millipedes and it may be uncertain which initiates the problem. Millipedes possess biting type mouthparts and gnaw at plant stems, shoots and roots usually

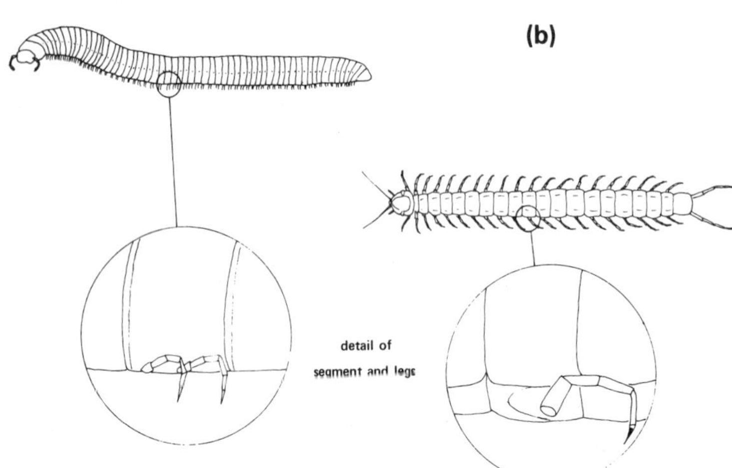

Fig. 19. The body form of (a) a millipede and (b) a centipede.

at about ground level. Seedlings and young tender plants are particularly susceptible to injury and seeds may also be directly attacked.

Eggs are laid in the soil and breeding is continuous under favourable conditions, though millipedes are quiescent during the winter. Development is extended and takes six months or more. The early stages are like a miniature version of the adult though there are fewer body segments and the colouration is much lighter. There are several moults during which further body segments are added before maturity is reached.

Symphilids

Symphilids are closely related to millipedes and centipedes (see Table 7) and in fact in North America are referred to as garden or white centipedes. This name is unfortunate as symphilids are plant feeding in contrast to true centipedes which are entirely predatory. They are however very like small white centipedes in appearance.

Full grown symphilids are about 6-8 mm in length, entirely white and with 14 body segments and 12 pairs of legs (Fig 20(a)). The body terminates in a pair of cerci. Symphilids inhabit the soil and move through cracks and crevices with great rapidity. They feed for the most part on the roots of plants, particularly the fine rootlets and root hairs. A great variety of plants may be attacked. Symptoms are stunted growth and discolouration of foliage due to impaired root action. Symphilid damage is particularly noticeable with transplants during the first few weeks after planting out, eg tomatoes,

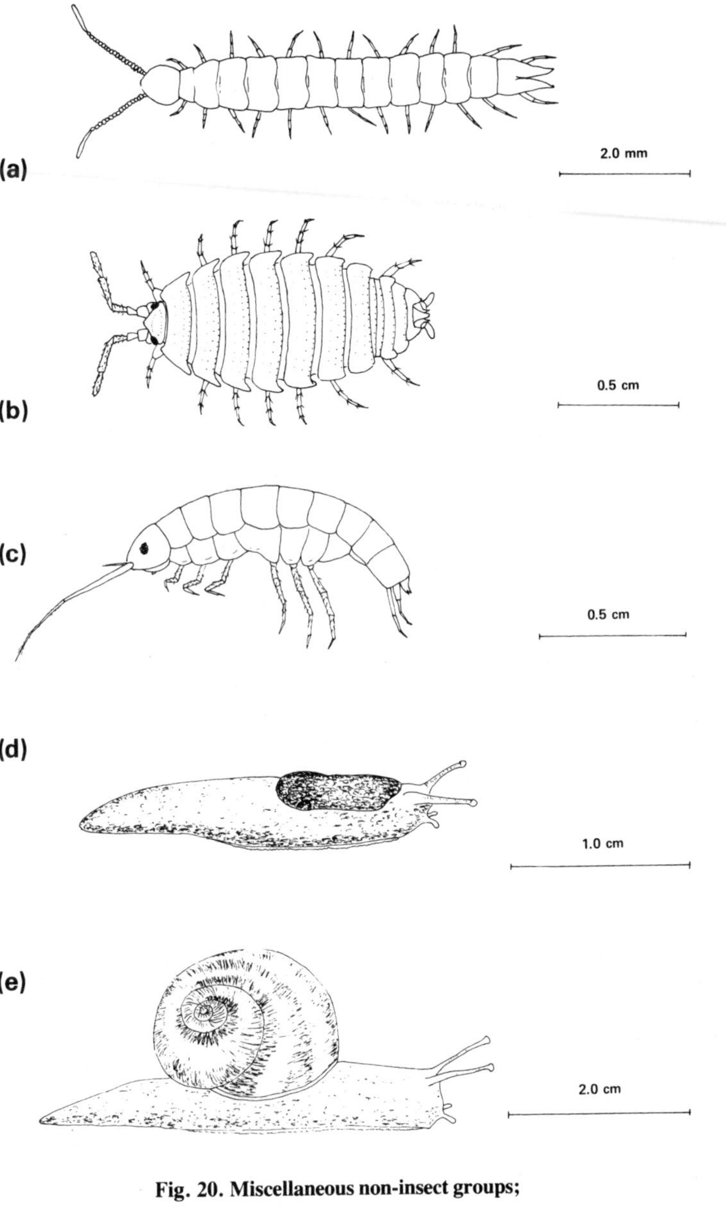

Fig. 20. Miscellaneous non-insect groups;

(a) symphilid,
(b) slater (woodlouse),
(c) amphipod (land shrimp),
(d) slug,
(e) snail.

and is more common under glass than out of doors. Symphilids appear to be favoured by highly organic soils. They are not easy to find as they move so rapidly. If symphilid injury is suspected a few plants should be lifted together with a ball of soil which must then be plunged immediately into a container of water. If the soil is gently broken apart under water any symphilids present will float as their body surface is highly water repellent.

Breeding is continuous under favourable conditions. Eggs are laid in the soil and development to adulthood takes about three months. There are many moults and moulting continues after maturity is reached. The immature stages are similar to the adults but initially have fewer body segments and thus fewer legs. Symphilids retreat deep into the soil in winter, a habit which makes them difficult to control completely.

Slaters (Woodlice)

Like millipedes, slaters feed primarily on dead and decaying plant remains and only occasionally attack growing plants. They belong in the class Crustacea of the arthropods (see Table 7) and are thus related to crabs and crayfish. Crustacea are for the most part aquatic (in fresh and salt water) and breathe by means of gills rather than tracheae. Slaters are one of the few groups of Crustacea that have become terrestrial in habit, but they retain gills for respiration and thus need a moist environment for survival.

In appearance slaters are greyish in colour, rounded on the upper surface but flattened underneath. They are about 1 cm in length when mature but some species are larger. The body is distinctly segmented with about ten segments, most of which bear a pair of legs. Some species tend to curl up into a ball when disturbed. Crustacea generally possess two pairs of antennae (as opposed to insects' single pair), but in slaters the second pair are very reduced in size so that effectively they have only a single pair. The general body form of a slater is shown in Fig 20(b).

Another form of terrestrial crustacean may sometimes be found in the same sort of situation as slaters. These are land shrimps (**Amphipods**) and are of no significance as pests. They are about the same size as slaters but brown in colour and flattened from side to side instead of from top to bottom. They possess two distinct pairs of antennae, one pair being much larger than the other (see Fig 20(c)).

When slaters injure plants they do so by chewing the base of stems and roots at and below soil level. They are mainly nocturnal in habit. Soft succulent plant parts are those mostly attacked. Slaters are favoured by conditions of high moisture, high organic matter and plenty of shelter such as that provided by piles of planks, stones and loose brickwork. Changing these conditions minimises the problem.

Eggs are produced in batches of 100-200 and are initially retained in a brood pouch on the underside of the female for about six weeks. On hatching from the egg the young are similar in form to the adults. The adults live for several years and may not breed until they are two years old. There are normally two broods a year. The winter is spent in an inactive state in sheltered places.

Slugs and snails

All the non-insect pest groups discussed so far are arthropods so that like insects, they possess an external jointed skeleton. Such however is not the case with slugs and snails. They are built to an entirely different body plan and are placed in a separate phylum of the animal kingdom, the **Mollusca** (see Table 7). Slugs and snails are entirely soft bodied except for the protective shell which is a prominent feature of snails. They are not segmented, again unlike arthropods. Two highly characteristic features of slugs and snails are the prominent flattened "foot" on which they move, and a rasp-like tongue (**radula**).

Slugs and snails can be serious pests of plants particularly in moist situations. They tend to feed on the softer parts of plants both above and below ground. Some species of slugs feed exclusively underground. Seedlings and young transplants are particularly vulnerable and seeds themselves may be attacked before germination is completed. Symptoms of injury are ragged holes in leaves, stems, tubers or other plant parts. Seeds and tubers may be largely hollowed out with only a small external hole. Slugs and snails are active mostly at night but can usually be found in the vicinity of damage hiding away in the daytime. In addition to the nature of plant injury, evidence of slugs and snails is provided by the silvery tracks of slime trails which are often still visible though dried out. Feeding activity is favoured by warm moist conditions but not by continuous rainfall. Snails hibernate over the winter but most slugs feed throughout the year when conditions are suitable and some species are active at temperatures down nearly to freezing point.

In some situations, eg pastures, damage from slugs may be much more important than is generally realised as high populations occur in many areas. Slug problems in the establishment of field crops are accentuated by minimum tillage techniques. These do not provide the soil disturbance of conventional cultivation which reduces slug populations.

The general appearance of slugs and snails (Figs 20(d) and (e)) is familiar to everyone and needs little description. The entire body is soft except for the inert shell. In snails this is large enough for the creature to retract completely inside when disturbed or under adverse conditions. The shell of slugs is very reduced or entirely absent. There is a saddle-like swelling on the back of slugs called the **mantle**. The flange-like foot secretes mucus over which the animal glides by wave-like contractions of its lower surface. A pair of long extensible tentacles bearing eyes arises from the head. A smaller pair of tentacles is located at the front end of the foot. The file-like radula, by which they rasp away plant tissue, is situated beneath the head of slugs and snails and is not readily visible externally.

Slugs and snails are **hermaphrodite**, that is both male and female reproductive organs are present in the one individual. Pairing usually still takes place though (to provide for the exchange of sperm), but self fertilisation can occur. Eggs are laid in clusters in the soil or other moist sheltered situations. They are gelatinous in texture, spherical and almost transparent. The eggs hatch to a small version of the adult, complete with delicate tiny shell in the case of snails. Growth is continuous without the need for moulting. The shell of snails is gradually enlarged as growth proceeds from young to adult form. The young stages of slugs and snails may feed more on dead and decaying plant remains than on live plants. They reach maturity after 3 to 12 months depending on the species, availability of food and environmental conditions. Snails may live for two years but slugs can have one or two generations a year. Eggs may be laid at any time of year but peak egg laying is usually in spring and autumn.

Slug and snail problems can be minimised by modifying environmental conditions so that they are no longer favourable, eg by removal of artificial shelter, reduction in organic manure, and improved drainage. Where additional measures are required, the usual treatment is with bran based baits incorporating either metaldehyde or the carbamate insecticide methiocarb.

SELECTED REFERENCES

Becker, P. 1974. *Pests of ornamental plants.* Bulletin No. 97. Ministry of Agriculture, Fisheries and Food. HMSO, London. 175 pp.

Evans, G. O., Sheals, J. G., Macfarlane, D. 1961. *The terrestrial acari of the British Isles*. British Museum, London. 219 pp.

Hughes, A. M. 1976, *The mites of stored food and houses*. 2nd ed. HMSO, London. 400 pp.

Hussey, N. W., Read, W. H., Hesling, J. J. 1969. *The pests of protected cultivation.* Edward Arnold, London. 404 pp.

Jeppson, L. R., Keifer, H. H., Baker. E. W. 1974. *Mites injurious to economic plants.* University of California. 614 pp.

Chapter 8

Insects and Plants

There are two very distinct types of association between insects and plants. Firstly, many insects utilise plants as a source of food; there is usually no benefit to the plant from the insects' feeding activities and the association is thus an antagonistic one. Over geological time plants have evolved various mechanisms to protect themselves from such insects and from larger herbivores. In the second type of association flowering plants exploit insects for the purpose of pollination and the insect at the same time obtains food in the form of nectar and pollen. Such an association is mutually beneficial. These two types of insect-plant interaction are considered separately in this chapter.

PLANT FEEDING INSECTS

The more important practical aspects of plant feeding (**phytophagous**) insects may be considered under four headings:

(a) Plant host range
(b) Types of insect injury to plants
(c) The relationship between injury and yield and quality of cultivated plants
(d) Insects and plant disease.

(a) Plant host range

The range of plants on which a given species of insect will feed is known as its **plant host range**. This may be narrow or wide but never extends to all plants. There is always discrimination to some extent. Three categories of phytophagous insects are usually recognised:

(1) *Monophagous insects* — are those that are confined to a single species of plant. Such an extreme habit is rare and few truly monophagous species exist. The classic example usually quoted is that of the silkworm which is practically confined to mulberry. Most so-called monophagous species of economically important insects in fact usually feed on a group of closely related plants.

(2) *Oligophagous insects* — characteristically feed on a group of botanically related plants usually within a single plant family. The habit is common amongst plant pests and examples include potato moth (*Phthorimaea operculella*)* which attacks only potato, tobacco and some other plants within the family Solanaceae, and diamond back moth (*Plutella xylostella*) which is confined to the cabbage family (*Cruciferae*).

(3) *Polyphagous insects* — are those that accept many plants from a diverse range of plant families. Preferences still exist however and some plants may be avoided altogether. In tropical countries some species of locusts have the reputation of being almost fully polyphagous.

Practical implications

There is obviously considerable practical importance in being able to define the plant host range for a pest species. Firstly, it indicates what other cultivated plants may be at risk in an area when an infestation develops on one kind of plant. In this respect it should be remembered that differences in plant susceptibility to a particular pest may extend to differences between cultivars as well as between species. This aspect is discussed further in Chapter 11 under pest control practices.

Secondly, where crop rotation is practised to assist with pest control information on the plant host ranges of pests is essential if this practice is to be of value. Finally, insects do not always confine themselves to cultivated plants. Suitable weed species may also be attacked and can serve as reservoirs of infestation, particularly in hedgerows or other uncultivated areas. Adequate knowledge of host plant ranges may allow such sources of infestation to be dealt with or at least taken into account when planning control measures.

Mechanisms of plant selection

Entomologists have been interested for many years in the mechanisms of host plant selection by insects and several theories were formerly put forward in explanation. One suggested that insects seek out and exploit those plants that provide the nutrients they require. This implies not only that different plants have different nutritional value but also that insects' requirements vary and that they can detect plant nutrient status and discriminate accordingly. An opposing theory was that all plants can provide all plant feeding insects with what they require nutritionally (as long as they eat them) and that discrimination is by means of secondary plant substances that act as attractants/repellents or feeding stimulants/deterrents. It is now known that neither theory is entirely true but that both contain some elements of truth.

* — a pest not established in Great Britain.

Table 11. Types of insect injury to plants and the main pest groups concerned

Mouthpart type	*Feeding behaviour*	*Type of injury*	*Main pest groups concerned*
biting and chewing	Remove solid portions of plant tissue from leaves, stems, flowers, fruits, or roots.	Holes or notches eaten in leaves or whole leaves consumed.	Larvae of most Lepidoptera; larvae and adults of some Coleoptera; some sawfly larvae (Hymenoptera); Orthoptera.
		One surface only of leaves eaten, leaving "windows".	Many immature larvae of Lepidoptera; larvae of diamond back moth; larvae of pear slug sawfly.
		Stems chewed externally at ground level.	Cutworms (larvae of certain Noctuid moths).
		Stems, shoots or branches bored internally.	Larvae of some Coleoptera, eg long-horn borers and a few Lepidoptera, eg currant clearwing.
		Developing seed consumed.	Larvae of some Lepidoptera, eg pea moth.
		Fruit tunnelled internally.	Larvae of some Lepidoptera, eg codling moth, plum fruit moth.
		Roots damaged.	Larvae of some Coleoptera, eg chafers.
		Surface of tubers chewed or tunnelled internally.	Larvae of some Coleoptera and of some Lepidoptera, eg potato moth.
true piercing and sucking	Feed on plant sap. May tap conducting vessels of plant. Inject saliva which may be toxic to the plant. Excrete honeydew. Usually attack only above ground parts of plants.	Weakening of plant through sap removal. Stunting, distortion, necrosis. Growth of sooty mould on honeydew.	Hemiptera* of all kinds.
superficial piercing and sucking	Pierce superficial cells only and remove contents. Attack mostly foliage and flowers.	Minute white flecks on attacked parts leading to bleaching or bronzing and premature death; may be accompanied by distortion.	Thysanoptera,* mites.*
rasping	Shred plant tissue. Ingest sap and macerated tissue. Usually bore or tunnel into attacked parts.	Ragged wounds to roots or stems; mines in leaves; tunnels in tap roots or bulbs.	Larvae of Diptera.

*Feeding by these groups may also involve transmission of plant viruses.

Considerable progress has been made in understanding the mechanisms of plant selection for some insects but others as yet have hardly been investigated. What is clear is that chemical composition of plants is of great importance in guiding the insect in the selection process. Volatile chemical constituents may attract or repel insects from a distance and, should contact be made, further discrimination may then occur due to other chemicals which stimulate or deter feeding or egg laying. The overall balance of such factors determines whether a plant is accepted or rejected. This is not to say that nutritional factors are not also important. Plants do vary in this respect and even if an insect feeds on a plant it cannot be regarded as a host plant unless it provides adequate nutrients and is without harmful toxicants. In addition to the chemical composition of plants certain physical factors such as leaf texture and toughness may also be important in determining whether a plant is accepted or not.

There is little doubt that the study of host plant selection has important practical implications. If we can understand the mechanisms by which an insect discriminates between host and non-host plants it may be possible to devise novel means of preventing damage to crop plants or to breed new varieties of plants which the pest will not attack. Unfortunately our understanding of these matters is not yet entirely adequate for any insect. However, this is a fertile field for further research that will surely yield important results.

(b) Types of insect injury

Insect injury to plants is caused almost entirely by their feeding activity, with the minor exception of a few insects such as cicadas which injure plants through their egg laying behaviour. The nature of plant injury is therefore determined primarily by the mouthpart structures of pests and by their feeding behaviour. This applies also to non-insect pests of plants such as mites, slugs, and snails.

The structure and functioning of the main types of mouthparts possessed by plant feeding insects have been considered in Chapter 3. The main types are listed in Table 11 together with brief descriptions of the feeding behaviour, type of injury caused, and the pest groups involved in each case. For the most part there is a clear distinction between insects with biting/chewing mouthparts which remove solid portions of plant tissue, and those with piercing/sucking mouthparts which feed on plant sap. Feeding in the former case results in obvious holes in leaves, stems, fruits or other plant parts, or tunnels bored in stems, fruits or roots. The actual nature of the injury is often characteristic of a particular pest. In contrast, insects with piercing/sucking mouthparts, which comprise the insect order Hemiptera, feed only on plant sap and are quite incapable of removing solid tissue. In the act of feeding however, these insects inject saliva into the plant to assist with digestion and uptake of sap. In many cases this saliva is irritant or toxic to the plant and results in stunting, distortion, or death of areas of plant tissue (**necrosis**). The effects of such insect feeding may thus be much more serious than the mere drain on the plant from sap removal. All insects which feed in this way also excrete **honeydew**. This is excess plant sap, high in soluble sugars, which is expelled in droplets from the hind end of the gut and which contaminates plant surfaces. Black sooty moulds grow on these sticky deposits. These fungi do not directly attack the plant but form an unsightly mess on the foliage of ornamental plants or on fruits (eg, tomatoes) which may then have to be washed before being marketed.

Because many piercing/sucking insects penetrate deeply into the plant with their stylets they provide an ideal means for the uptake and injection of many plant viruses. In some situations this aspect is much more important than any direct effects produced by the insects themselves, as for instance with aphids on potatoes. Insects as virus vectors are further discussed later in this Chapter.

Two other types of mouthparts and feeding behaviour need to be distinguished besides the major ones already referred to. These are the superficial piercing/sucking type and the rasping type (see Table 11).

Thrips (order Thysanoptera) and plant feeding mites possess mouthparts which are like an abbreviated form of the true piercing/sucking pattern. Stylets are present but they are much shorter than in the Hemiptera. Thrips and mites therefore are able to pierce and remove cell contents (including chlorophyll) only from superficial cells of the plant. This usually results in minute white flecks appearing which eventually coalesce leading to a bleached or bronzed appearance. Mites and thrips also inject saliva as they feed and with some species this produces marked distortion of the plant parts attacked. Such is particularly the case with eriophyid mites whose feeding may lead to the production of distinct plant galls. Some mites and thrips can act as vectors of certain plant viruses but as a group they are less important in this respect than hemipterous insects because their mouthparts do not penetrate so deeply into plants.

The final plant feeding mouthpart type, referred to as rasping, is that possessed by larvae of most Diptera (flies). It consists of a pair of sharp recurved hooks which protrude from the mouth and which are used to shred plant tissue (see Fig 7). The resultant sap and particles of material are then ingested. Many larvae which feed in this way tunnel into (mine) the parts of the plant attacked.

Besides the discrimination which insects exhibit as to species of plants attacked they are also usually very selective as to the part of the plant on which they feed. This, together with the mode of feeding for a particular pest, may result in plant damage symptoms that are highly characteristic for the insect concerned. For example pea pods with the peas partly consumed and sawdust-like excreta present within the pod can confidently be ascribed to pea moth as no other pest causes similar injury in Great Britain. On the other hand root injury to plants could be caused by numerous insects such as chafer grubs, larvae of vine weevil or several other weevil species, wireworms, or larvae of swift moths, depending on the locality, the kind of plant concerned and the time of year. A detailed breakdown of the kinds of plant injury caused by insects (and mites) is given in column three of Table 11.

(c) Relationship of pest injury to yield and quality of produce

Because the farmer or grower is basically interested in the effects of pest attack on quantity and quality of his harvest, it might be expected that this would have received detailed study. In fact, it is a rather neglected area of research which is only now beginning to receive adequate attention. However, sufficient is known to appreciate that the interaction between pest injury and crop yield is complex with a number of important components that may influence the outcome. Five major kinds of factors may be identified:

(1) Nature of the injury;
(2) Part of the plant attacked in relation to that which gives rise to yield;
(3) Intensity of the injury;
(4) The time when injury occurs in relation to growth of the plant;
and
(5) Effect of environmental conditions on the ability of the plant to withstand injury.

(1) *Nature of the injury*

This has been discussed in detail in the preceding section in which types of insect mouthparts and feeding behaviour were considered. No further elaboration is necessary here.

(2) *Part of the plant attacked*

We have seen that the part of a plant attacked varies with the pest concerned. The effect of any injury moreover depends to a great extent on which part of the plant is harvested by man. This varies greatly depending on the particular plant. In some cases it is the leaves that are the desired part (lettuce, spinach) but with other plants it may be roots (carrots, radishes) or tubers (potatoes); in other cases fruits (tomatoes, apples), seeds (peas, cereals) or for a few plants leaf stalks (celery, rhubarb) may be chosen.

The same kind of injury to different plants may thus result in quite different effects on yield, depending on the part of the plant that is harvested. We can distinguish between injury to **yield forming organs**, which may be described as *direct*, and injury to **non-yield forming organs**, where the effect on final yield is *indirect*. The importance of these considerations is that generally plants can withstand more injury to non-yield forming parts (indirect injury) than they can to yield forming organs (direct injury). This is discussed further under (3), intensity of injury.

The foregoing discussion applies to cultivated plants from which only a certain part is harvested and which constitutes the yield. This category includes cultivated plants which man utilises for food, fibre, or some other specific product. However, plants are also grown for decorative purposes, either as permanent landscape plantings or as short term flower crops or pot plants. In these cases the "yield" is not readily definable as it may constitute all of the above ground parts of the plant. The effects of pests in such cases must be considered in terms of "quality" of product; that is possible downgrading of aesthetic appearance and reduced vigour or rate of growth. The concepts of yield and non-yield forming organs thus have little application when dealing with ornamental plants.

Quality of produce is also of major importance in relation to market value of many horticultural food crops so that the effects of pest injury must then be considered in these terms as well as in terms of gross yield. Some pests do not significantly reduce the weight of produce harvested but quality may be seriously downgraded so that the market value is severely affected. In extreme cases the produce is rendered unsaleable. Wireworm injury to potato tubers for example may not seriously affect gross yield but quality is drastically downgraded.

(3) *Intensity of injury*

The degree of injury to a plant is obviously closely related to the density of the pest population (the more insects per plant, the greater the degree of injury will tend to be) but the stage of development of the pest is also important. Large larvae eat more than smaller ones and adult insects may not feed at all, so these complications must be taken into account. It is thus perhaps better to think in terms of intensity of plant injury rather than in pest population density, though in most situations the two vary together.

Generally, the greater the degree of injury to a plant the lower the yield will be but the end result depends very much on whether yield forming or non-yield forming parts of the plant are attacked. Where injury is to non-yield forming organs the plant often has considerable capacity to absorb the effects of injury either because it has an excess of the tissue concerned or because it can make compensatory growth to replace the injured parts before it reaches maturity. The general pattern therefore is that low levels of indirect injury to plants have no effect on final yield so that if we plot yield against increasing pest injury (Fig 21(a)) the first part of the graph is a level plateau. With higher levels of injury, however, the point is reached where the plant can no longer withstand the effect and yield starts to decrease. This is the point in Fig 21(a) where the curve just starts to flex downwards. This is known as the **damage threshold** and is a very important concept because below it pest populations are of no significance whereas above it increasing depression of yield occurs. This topic is referred to again in Chapter 11 where pest control procedures are discussed. As an example, the damage threshold for glasshouse cucumbers from mite injury has been determined to be about 30 percent loss of foliage.

Over a range of injury levels above the damage threshold, yield progressively decreases with increasing pest injury (see the linear part of the graph in Fig. 21(a)) but with higher levels of injury still, the curve of yield flattens out so that 100 percent loss does not usually occur even with the highest pest populations. This is perhaps because at high pest densities interference between individuals may occur and closely adjacent injuries may exert less effect than ones widely spaced. One further point should be noted from Fig. 21(a). Low levels of pest injury (to non-yield forming organs) may sometimes actually stimulate greater yield than the total absence of the pest (dotted part of the curve). This has been shown to occur for instance with some foliage pests of potatoes. It appears to be an anomaly but may be due to some sort of pruning effect and points up the complexity of plant reaction to injury. Fig. 21(a) is a generalised graph depicting what is believed to be the pattern for most pests which cause indirect injury to plants. In few specific cases however is information adequate to draw a detailed picture, and often data is only available over an intermediate range of pest densities covering the linear (straight line) part of the graph.

In contrast, pests which cause injury directly to yield forming organs of plants exhibit a very different relationship between pest density and yield. The damage threshold in such cases is extremely low (or in fact zero) and yield decreases in linear fashion until, with

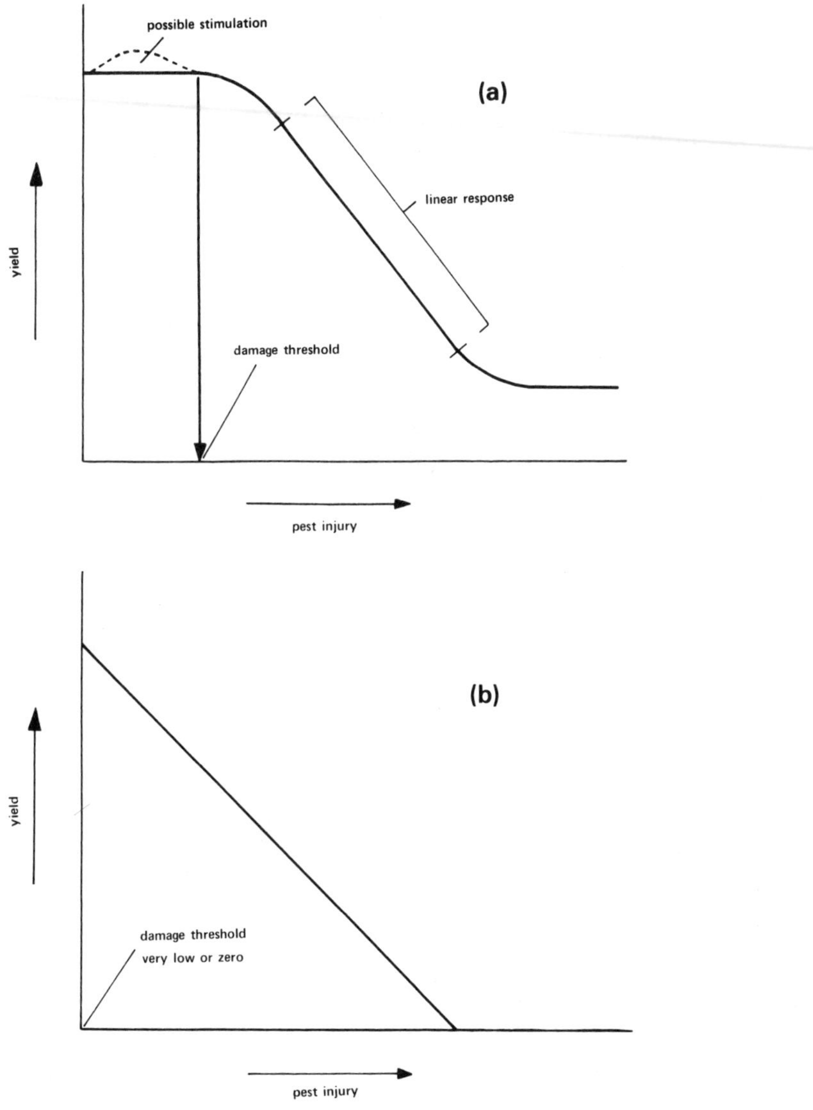

Fig. 21. Generalised curves of relationship between yield and pest injury where (a) injury is to non-yield forming organ, ie, indirect, and (b) injury is to yield forming organ, ie, direct.

high pest populations, it reaches zero. Such a relationship is depicted in Fig. 21(b). Codling moth is an example of such a pest. Virtually every codling moth larva can account for an apple fruit and uncontrolled populations may cause complete loss of marketable crop.

Indirect production systems involving conversion of plant material to animal matter (livestock farming), add a further dimension to the problem of defining the significance of plant pest injury and of relating injury levels to yield. In such cases plant material has no monetary value until it is converted into animal matter and furthermore, complete utilisation of pasture growth is rarely achieved. Reductions in plant growth brought about by pest attack therefore only become significant when they coincide with livestock feed shortages. Effects of pests in pastures are thus of greater practical significance where stocking rates are high and where there is high utilisation of plant growth than where the reverse applies. Put another way, we can say that the grazing animal provides a considerable buffer between pest injury to the plant and the final harvested yield in the form of meat, wool or dairy produce. This is not to say that pasture pests are not important but simply that their effects are much more difficult to quantify compared to field or horticultural crops.

(4) *The time when injury occurs*

For individual plants the effect of pest injury varies according to their stage of growth. Generally, injury which occurs early in the life of a plant is less likely to have serious effects than injury inflicted later because the plant has more chance to recover. For most plants four main stages in their growth cycle can be recognised which differ in susceptibility to injury.

(i) *Seedlings* — Young, newly emerged seedlings are usually very susceptible to injury. One bite from an insect with biting/chewing mouthparts or one puncture from a piercing/sucking insect may mean the death of a plant. New transplants are also vulnerable before they have become properly established, but their tolerance level is higher than seedlings because they are larger. Plants which are vegetatively propagated, such as potatoes, avoid the extremely sensitive seedling stage.

Loss of whole plants in many crops at the seedling stage may have little or no effect on final yield until a high percentage of the plant population has been destroyed. The reason is that many field crops, such as cereals, are grown at high population densities rather than as spaced plants and, because of this close planting, individual potential plant yield is never achieved. If some loss of plant stand takes place at the seedling stage the remainder of the plant population may well be able to compensate by harvest time. Thus it has been shown for example that sugar beet at a planting density of 30,000 per acre (75,000/ha) can lose 50 percent of plants at the seedling stage without final yield being reduced.

(ii) *Young plants* — Once a seedling is well established it enters a phase of fast and vigorous growth. Because of this, and its ability to make compensatory growth, the young plant is quite tolerant of pest injury providing this is not to vital plant parts that cannot be replaced.

(iii) *Stage of formation of yield forming organs* — When the yield forming organs of a plant begin to be laid down susceptibility to pest injury rises again, particularly for those pests that directly attack the yield forming parts. For many plants the formation of yield forming organs takes place only once in their growth cycle so that if they are destroyed they cannot be replaced. Pip and stone fruits for instance have only a short flowering season following which the young fruitlets develop. If these are damaged or destroyed no compensatory growth can make up for it as they are not replaced. More fruits may be set of course than the plant can mature, so that if they are thinned as a result of pest attack there may be no ill effect on final harvest, but pests are not usually so convenient in their habits.

(iv) *Mature plants close to harvest* — Once plants are mature and the yield forming process has reached completion the effects of pest injury again become much less important, unless of course the part of the plant to be harvested is directly attacked. Thus foliage injury to potatoes once the tubers are fully formed is of no consequence, but if pests such as wireworms attack the tubers directly this is important. A further point is that for some cultivated plants the harvestable portion itself becomes less attractive to pests as it matures. This is especially true for cereal grains which at maturity are too hard and dry for most pests to tackle. The four stages of a cereal plant with respect to sensitivity to pest injury are shown pictorially in Fig. 22.

These comments apply for the most part to plants with an annual growth cycle. Perennial plants (eg perennial fruits) do not fully fit the pattern. They avoid the very susceptible seedling stage (except for nursery propagation) and injury to foliage or roots may produce effects the season *after* it occurs by affecting such processes as bud formation.

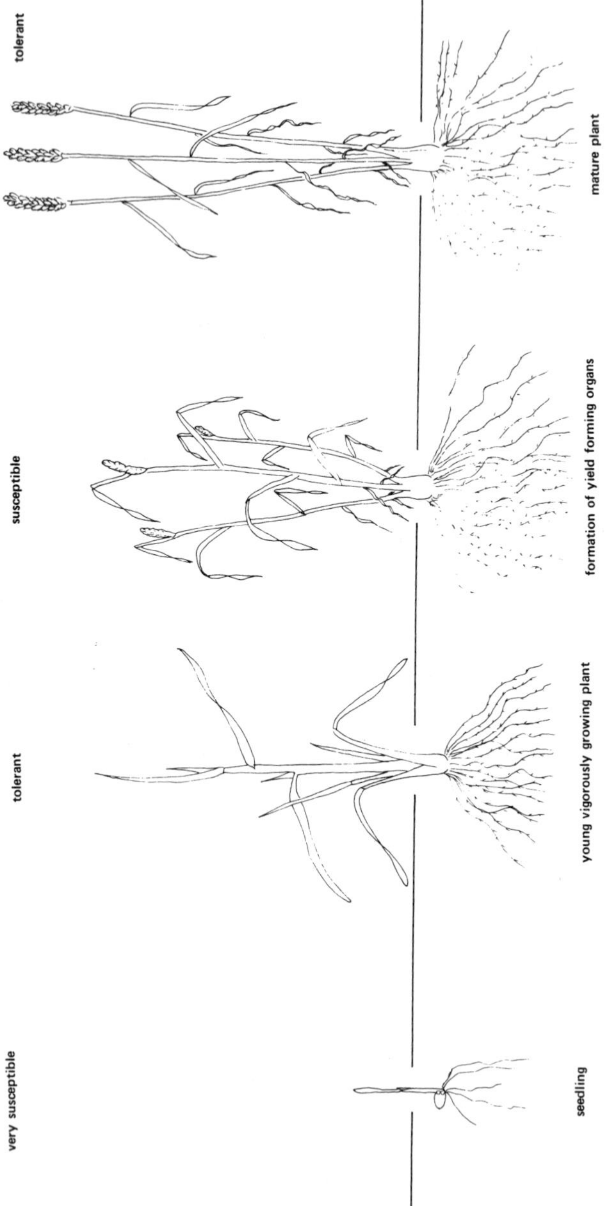

Fig. 22. The susceptibility of a cereal to pest injury at different stages of growth.

(5) *The effect of environmental conditions*

It has sometimes been suggested that plants which are growing poorly and perhaps showing signs of mineral deficiency are more heavily attacked by pests than plants which are not stressed in this way. There is in fact very little evidence to support this, with the possible exception of some sap feeding insects and mites. What is certain however is that plants which are nutrient deficient or under moisture stress have less ability to withstand pest injury and to make compensatory growth. It is good growing practice therefore to ensure that cultivated plants are well watered and fed both to obtain maximum yields and to minimise the effects of pest injury.

Conclusion

It should be clear from the foregoing discussion that the relationship between pest injury and yield of cultivated plants is complex with many interacting factors. The outcome in terms of yield and quality of produce can therefore be expected to vary according to circumstances. That such is the case in practice is clear but unfortunately in few situations do we have sufficient information to understand the process in detail or to predict accurately the outcome of different levels of pest attack. Nevertheless the concepts discussed are important ones, particularly that of the damage threshold level for pests below which they do not present a problem. This has great significance in relation to pest control and is discussed further in Chapter 11.

(d) Insects and plant disease

If plant disease is defined as a condition of ill health leading to reduced growth, lowered production or even death of the plant, then some insects can be regarded as the direct cause. This is true for those that feed on plant sap resulting in stunting and distortion without obvious loss of tissue. Such a category of plant disease is recognised by some authors (Carter 1973). For the most part however, plant diseases are induced by micro-organisms the most important of which are fungi, bacteria and viruses. In many cases insects appear to play no part in disease development but for some plant diseases insects are important in assisting spread or in facilitating the infection process. The extreme is reached by certain plant viruses (such as barley yellow dwarf virus and pea leaf roll virus) which are entirely dependent on particular insects for their transmission. Such insects are known as **vectors** for the micro-organisms concerned and present the possibility of controlling the disease by controlling the vector in a similar way to some insect-borne human diseases such as malaria.

The degree of association between insects and plant pathogenic micro-organisms ranges from completely accidental and casual to fully obligatory, at least as far as the micro-organism is concerned. Three main categories may be recognised:

(1) *Casual associations*

Most insects are well endowed with body hairs and spines and thus easily become contaminated with fungal spores and bacteria. By their activity the insects can then carry infection from one place to another. The causal organism of fireblight of pip fruits and certain rosaceous ornamentals (*Erwinia amylovora*) is an example of a plant disease organism distributed in this way. The association seems to be entirely accidental and the insect does not benefit in any way.

Besides distributing micro-organisms in this way, the feeding activities of insects create wounds in plants which facilitate entry and initiation of disease infection. This applies to micro-organisms that may already be present on the plant surface as well as to those that have been carried by the insect. Not all plant wounds result in disease infection though because of natural resistance mechanisms. Insect injury seems to be particularly important in facilitating infection of ripening fruits, eg by *Monilinia fructicola* the causal organism of brown rot, and of plant roots, in the latter case perhaps because the soil is such a rich microbiological environment.

(2) *Regular symbiotic associations*

In these cases we find a certain micro-organism regularly associated with a particular species of insect which not only assists in its spread but also helps to provide the right conditions for it to develop. In return the insect usually feeds on the micro-organism so that the association is mutually beneficial (**symbiotic**). Examples occur among various bark beetles which are usually associated with particular species of fungi. The fungus concerned may be a serious plant pathogen as for example *Ceratocystis ulmi*, the cause of Dutch elm disease. In such cases the two organisms involved can exist apart but do not thrive and are usually found together.

(3) *Intimate obligatory associations*

In this type of association the micro-organism is entirely dependent on the insect for transmission from plant to plant. The main examples are many of the plant viruses. It is uncertain whether the insect derives any benefit from the association but on the other hand it does not suffer harmful effects. Nearly all the insects involved in this way are from the order Hemiptera, in particular aphids and leafhoppers, but some species of thrips and beetles are also implicated. Some mites and nematodes are also known to transmit certain plant viruses. Insects with biting mouthparts such as beetles are comparatively rare as plant virus vectors and also are rather inefficient.

Two main insect/plant virus associations are recognised in aphids, which are the main virus vectors in temperate climates. These are the **non-persistent** or **stylet-borne** type and the **persistent** or **circulative** type (see Table 12). With non-persistent viruses external contamination of the insect's mouthparts occurs and transmission from infected to healthy plants is mainly a matter of mechanical transfer. The insect acquires the virus after only a minute or so of feeding and can transmit it immediately but it does not remain infective for long. In contrast, persistent viruses require several minutes or even hours of feeding before the insect becomes infective and there is a period of delay before the virus can be transmitted. Once infected the vector is capable of transmission for

Table 12. The major types of insect/plant virus associations and their characteristics

Type	*Characteristics*
non-persistent (stylet-borne)	Rapid acquisition (seconds to minutes of insect feeding). Brief retention time (few minutes to a few hours). Immediately transmissible. Not amenable to control by systemic insecticides.
Example	Potato virus Y, vector — peach — potato aphid (*Myzus persicae*)
persistent (circulative)	Slow acquisition (minutes to hours of insect feeding). Long retention time (days to weeks). Not immediately transmissible — "incubation" period required. Amenable to control by systemic insecticides.
Example	Barley yellow dwarf virus, vector — bird-cherry aphid (*Rhopalosiphum padi*).*

Note — Individual viruses require specific insect vectors, but one virus may be transmitted by several vectors, and one species of vector may be capable of transmitting several viruses.

* — Some strains of BYDV are transmitted more efficiently by other species of aphids.

a long time, even for its whole life span in some cases. Persistent viruses pass into the insect's circulatory system (hence the alternative name of circulative viruses) and eventually enter the salivary glands so that when the insect feeds and injects saliva into a plant the virus particles are injected with it. In a few cases the virus may actually multiply within the aphid's body during this process. The major characteristics of these two types of insect/virus associations are set out in Table 12.

As insect borne plant viruses are not normally transmitted other than through the appropriate vector there is the possibility of controlling their spread by control of the vector. With aphids there are two possibilities. Either plants can be grown in an area or at a time when aphid populations and activity are low, or insecticides can be used to suppress aphid numbers. The former conditions can be satisfied in localities which are elevated (so that temperatures are low), and exposed. This is the basis for designating certain areas for potato seed production for instance. There is also the possibility of adjusting planting dates to avoid peak aphid populations at critical times. Thus winter wheat sown in late autumn will to a considerable extent emerge after peak aphid populations have declined and thereby avoid heavy infestation. In practice however the trend is to earlier planting as this can lead to enhanced yields.

Control of aphid-borne viruses by systemic insecticides is for the most part only effective against persistent viruses as the non-persistent ones can be transmitted by a feeding aphid before it takes up the insecticide and is killed. Nevertheless there is a useful place for this kind of treatment with certain crops such as potatoes which are affected by several virus diseases. The usual form of treatment is to apply a granular preparation of a systemic insecticide to the soil at planting time. This protects the plants for the first six to eight weeks of life. Additional later applications in spray form may be desirable. Anyone considering attempting to control an insect-borne plant virus by means of insecticides applied against the insect vector should seek detailed advice.

POLLINATING INSECTS

Pollination is the transfer of pollen from the male organs of flowering plants (**anthers**) to the female receptive part of the flower (**stigma**). It is essential for the fertilisation of the ovule and consequent seed set and development. Effective pollination is therefore required for normal production of crop plants where the harvested part of the plant consists of seeds or fruits. Furthermore, pollination is a necessary requirement for all plants that are propagated by way of seed.

Some plants are able to undergo **self pollination**, that is transfer of pollen from anthers to stigma within the same flower. With a few crops such as peas, this provides for adequate seed set but for most plants self pollination results in poor seed yields. **Cross pollination** (transfer of pollen from other plants of the same species) must therefore normally take place. According to the mechanism of cross pollination, so plants can be divided into two major groups:

(a) those that are wind pollinated, such as grasses, and
(b) those that are pollinated by insects

Plants which are wind pollinated usually have inconspicuous flowers and large amounts of pollen are produced to ensure that some reaches its target. The pollen grains are very small, light, and easily dispersed. Insect pollinated plants on the other hand usually have larger conspicuous flowers which are often of distinctive form, brightly coloured and scented. Moreover, special glandular structures (**nectaries**) which secrete a sweet sugary solution (**nectar**) are usually located within the flowers and sometimes on other parts of the plant. These are clearly features that have evolved to attract insects for the purpose of pollination and it is true to say that without pollinating insects there would not be the wonderful array of flowers which exist. Pollen from insect pollinated plants tends to be sticky so that it adheres to the bodies of insects when they probe into the flower.

The importance of ensuring adequate pollination of fruit crops for maximum production has been recognised by fruit growers for many years and bees are often managed or hired for this purpose. With some fruit crops such as kiwifruit,* inadequate pollination may limit fruit size but the practical solution in this case is not simple as kiwifruit flowers provide little reward for visiting bees. Seed production of certain plants can also be severely limited by inadequate pollination. A prime example is lucerne. In a very few instances insect pollination of a crop plant is undesirable, as in the case of glasshouse cucumbers where it leads to swollen and poor quality fruit.

Bees, of which there are a number of species, are the main insect pollinating group and are by far the most important in practice. However, many other insects such as butterflies, moths and some flies and beetles do visit flowers quite frequently and may play a part in pollinating some plants. The main reasons why bees are so important as pollinators are that (a) they consistently, over a period of time, visit flowers of the same species, and (b) their hairy bodies readily pick up and carry a pollen load. Although a large number of species of (mainly solitary†) bees occur in Great Britain they are for

* — *Actinidia chinensis*

† — Solitary species are those in which pairs of bees raise their own offspring, as opposed to social species which live in complex colonies with a single reproductive queen.

the most part not well adapted to the pollination of cultivated plants. The most important pollinators are the honey bee and to a lesser extent (but particularly for certain crops) the bumble bees. Besides their better adaptation to crop plants, honey bees have the big advantage that they can be manipulated to provide improved pollination where it is needed.

(a) Honey bees

Man's experience with honey bees spans several thousand years and the honey bee is one of the very few "domesticated" insects. Although much is known about bees from a practical point of view, their behaviour and colony organisation are so complex that much new information continues to be discovered. Only brief reference to certain aspects is made here. The production of honey is the main reason for the existence of the British beekeeping industry and manipulation of honey bees for pollination is to a large extent secondary. This contrasts with some other parts of the world, North America in particular, where much of the beekeeper's income may be derived from renting hives to fruit growers for pollination purposes.

Only one species of honey bee (*Apis mellifera*) occurs in temperate regions of the world but there are many races which may differ in vigour and foraging behaviour. Each colony (hive) is a complex society of up to 50-60,000 individuals. There is normally one reproductive female (**queen**) per colony. She may live for several years but commercial beekeepers normally renew the queen every other year to maintain the vigour of the hive. If the queen dies another will develop to take her place. Most of the other colony members are **workers** (sterile females) which forage, collect nectar and pollen as food, tend the larvae within the hive, construct new cells and perform any other tasks required. Male bees (**drones**) are only produced at certain times. They take little or no part in running the colony, their prime function being to mate with potential future queens. After mating, the queen can regulate fertilisation of eggs as they are laid. Unfertilised eggs give rise to drones, fertilised eggs to workers and queens. Whether a fertilised egg develops into a queen or a worker depends on the nature and amount of food provided to the larva. Those larvae destined to produce queens are housed in special large cells.

Because the queen is replaced if she dies, bee colonies can continue indefinitely. From time to time a proportion of the workers, together with the old queen, leave the nest en masse and move some distance to settle elsewhere, a process known as **swarming**. The factors which trigger swarming behaviour are not well understood.

Provision for a new queen to take the place of the old one has been made in the parent colony prior to departure of the swarm. Swarming is not popular with beekeepers because colonies which swarm lose up to half of their population and therefore do not gather a large honey crop.

Honey bees are particularly valuable as pollinators of cultivated plants for several reasons:

(1) they will visit a very wide range of plants;
(2) but at any one time a bee tends to visit only one kind of plant; this behaviour is particularly valuable for transfer of pollen between plants of the same species;
(3) hives can be moved to concentrate numbers of bees at the place and time they are most needed.

There are limitations though on how well bees will work a particular crop if nectar production is low and there is an easier source near by. Honey bees are also limited in their activity by weather conditions. They do not forage to any extent below 15°C and are grounded by strong winds. Thus some plants which flower early in the spring, for example some varieties of plums, may not be adequately pollinated by honey bees.

Another problem is that the structure of some flowers (for example those with deep corolla tubes with nectaries at the base) is such that honey bees will not frequent them. Accordingly other species of bees may need to be considered for certain crops.

(b) Bumble bees

More than a dozen species of bumble bees (mostly *Bombus* spp.) occur in Great Britain. Some are short-tongued types, others long-tongued. Bumble bees make small annual nests and maintain themselves naturally but are never very numerous. The long-tongued species are good pollinators of lucerne, red clover and some other legumes but are not usually present in sufficient numbers for effective pollination of large areas. Besides their robust build and long tongues compared to honey bees, bumble bees also forage at lower temperatures (down to about 5°C). Bumble bees are not managed to any extent but knowledge is available to permit greater use of them if it were desired.

SELECTED REFERENCES

Anon. 1981. *Beekeeping*. Bulletin No. 9. MAFF. 11th edition. HMSO, London. 25 pp.

Bardner, R.; Fletcher, K. E. 1974. Insect infestations and their effects on the growth and yield of field crops: a review. *Bulletin of Entomological Research*. 64. 141-160.

Butler, C. G. 1976. *The world of the honey bee*. Collins, London. 226 pp.

Carter, W. 1973. *Insects in relation to plant disease*. Wiley, New York. 759 pp.

Crane, E. 1975. *Honey: a comprehensive survey*. Heinemann, London. 608 pp.

Dethier, V. G. 1970. Some general considerations of insects' responses to the chemicals in food plants. In "*Control of insect behaviour by natural products*" D. L. Wood, R. M. Silverstein, M. Nakajima. eds. pp 21-28. Academic Press, New York. 345 pp.

Free, J. B. 1970. *Insect pollination of crops*. Academic Press, London. 544 pp.

Free, J. B. 1977. *The social organisation of honey bees*. Studies in Biology No 81. Arnold, London. 64 pp.

Gojmerac, W. L. 1980. *Bees, beekeeping, honey and pollination*. A. V. I. Westport, Conn. 208 pp.

Harborne, J. B. 1977. *Introduction to ecological biochemistry*. Academic Press, London. 243 pp.

Hussey, N. W.; Parr, W. J. 1963. The effect of glasshouse red spider mite (*Tetranychus urticae* Koch.) on the yield of cucumbers. *Journal of Horticultural Science*. 38. 255-263.

Jones, F. G. W.; Dunning, R. A.; Humphries, K. P. 1955. The effects of defoliation and loss of stand upon yield of sugar beet. *Annals of Applied Biology* 43. 63-70.

Martin, E. C.; McGregor, S. E. 1973. Changing trends in insect pollination of crops. *Annual Review of Entomology*. 18. 207–226.

Richards, A. J. 1978. *The pollination of flowers by insects*. Linnaen Society Symposium No 6. Academic Press, London. 214 pp.

Todd, F. E.; McGregor, S. E. 1960. The use of honey bees in the production of crops. *Annual Review of Entomology*. 5. 265-278.

Chapter 9

Predators, Parasites and Pathogens

The high potential rate of increase of insects is counterbalanced most of the time by high mortality. There are two broad groups of natural mortality factors, those that are associated with physical aspects of the environment and those that are due to the activities of other living organisms. Of the latter, other insects, in the form of **predators** and **parasites**, are among the most significant but insects also suffer from diseases caused by micro-organisms (**pathogens**) and in some situations these may be extremely important. All of these **natural enemies** of insects have the potential for manipulation for control of pest species. This is the field of **biological control** which is discussed in Chapter 11. In this section the groups of organisms concerned and their biology are described.

Predators

Predators of insects are animals which capture and consume them as a source of food. Many are other insects (or at least other arthropods), but some other animals also feed on insects, either exclusively or as one component of a more mixed diet. These include some vertebrate animals such as certain birds and small mammals.

(a) *The main groups of insect predators and their prey*

Table 13 lists the main groups of insect predators and other animals which prey on insects. The predatory habit is quite widespread among different insect groups and some predatory species exist in most orders. Only the more important groups are listed in the Table. Several families of beetles (Coleoptera) are largely predatory in behaviour. Amongst these the ladybirds (Coccinellidae) are particularly important as they attack mainly aphids, scale insects and mealy bugs, many of which are serious pests of plants. The ground beetles (Carabidae), rove beetles (Staphylinidae), and tiger beetles (Cicindelidae) are predators of a diverse range of ground dwelling insects.

A few orders of insects, such as the Neuroptera (lacewings and relatives), are exclusively predatory. In the orders Diptera (flies) and Hymenoptera (ants, bees and wasps) however, only certain families or individual species are predatory. The same applies to the Hemiptera where predatory species occur within several families of the sub-order Heteroptera.

Spiders, which are close relatives of insects, are exclusively predatory. Their prey consists mostly of insects which they capture by various means such as entanglement in their elaborate webs. The predatory habit is also quite common amongst mites whose prey consists mainly of other mite species. Such predatory mites are now being managed to provide control of certain plant feeding species.

The only vertebrates included in Table 13 are the birds. Other vertebrate animals such as field mice, shrews and bats feed heavily on insects but are of little significance for control of economically important species. Other higher animals such as lizards and toads may be quite important insect predators in some warmer countries.

(b) *Feeding behaviour of predatory insects*

Predatory insects are of two main types: those that actively seek out and capture their prey, and those that lie in wait and grab prey which come within range. The former are often highly active with prominent legs, eg adult tiger beetles. The latter may be equipped with special capturing devices such as the folding forelegs of mantids, which can be extended rapidly. Some predatory insects (eg ladybirds) show no special structural modifications as their prey are largely immobile and are simply "grazed".

The list of predatory groups in Table 13 includes examples of insects with biting/chewing mouthparts (beetles) or piercing/sucking mouthparts (bugs and mites). In the former the prey may simply be chewed to pieces, but quite often the predator's mandibles are drawn out into long sickle-shaped structures which are grooved along their inner surfaces, or even completely hollow as in lacewing larvae. In use, the tips of such mandibles are thrust into the prey and the fluid body contents sucked out, the empty skin of the prey being discarded when the process is complete. Predatory bugs and mites feed in a similar fashion to their plant feeding counterparts by piercing the outer body wall of their prey and tapping the body juices.

Many insects are predatory both as larvae and as adults. This is the case with ladybirds and lacewings. However with some other groups only the larvae or only the adults are predatory. For example, hover flies are predatory only as larvae and the adults feed primarily on nectar or pollen. Conversely, robber flies have predatory adults whilst the larvae of most species feed as scavengers amongst decaying plant remains.

(c) *Range of prey attacked*

The range of species on which most predatory insects will feed is not rigidly restricted so that composition of their prey may vary with place and time according to what is available. The same applies to an even greater extent to vertebrate predators. For many birds for instance, the size and location of insect food is probably more important than the actual species. Some predators however are much more selective as to the kinds of insects on which they will feed. Certain ladybirds for example will only accept particular species of aphids while others may be confined to one species of scale insect. In general, predatory insects are more flexible in their feeding habits than parasitic insects.

Parasites

Insect parasites (of other insects) are those whose larvae feed internally or externally on the body of another insect. The attacked insect is referred to as the **host** and sustains the parasite larva (or larvae) throughout its development. Insect parasites are invariably smaller than their hosts, only slightly so if a single parasitic larva develops within each individual host insect but much smaller where many parasite larvae develop, as is quite common. This contrasts with predatory insects which are always larger than their prey. Another point of distinction between parasites and predators is that only one individual host is required to support development of a parasite whereas predators consume several (often many) prey during their lifetime.

Insect parasites invariably kill their host by the time their own development is complete, which contrasts strongly with insect and other parasites of higher animals, such as fleas and lice. In the latter case the effect is more one of debilitation and annoyance rather than death. This contrast is so marked that some entomologists use the term **parasitoid** to refer to insect parasites of other insects to make the distinction clear, but the term is not in common use.

Parasitism of insects is extremely common and most insects have one, or often several species of parasites associated with them. However, the level of parasitism at any one time varies greatly with conditions.

(a) *The main groups of parasitic insects*

The parasitic mode of life is not nearly as widespread amongst insect groups as is the predatory habit and is for all practical purposes confined to the orders Hymenoptera and Diptera. However, within these two orders, particularly the Hymenoptera, large numbers of species are involved and almost as many parasitic insects may exist as free living ones. Only those that are of economic importance as natural enemies of pests have been studied to any extent so that our knowledge of how many actually exist is rather poor.

The main families within the Hymenoptera and Diptera which are parasitic, or predominantly so, are listed in Table 14 together with their principal hosts in each case.

Table 13. The major groups of insect predators and their prey

Group	*Stage which is predatory*	*Main prey*
Insects		
Coleoptera		
Coccinellidae (ladybird beetles)	Larvae and adults	Aphids, scale insects, mealy bugs.
Carabidae (ground beetles)	Larvae and adults.	Various soil dwelling insects.
Staphylinidae (rove beetles)	Larvae and adults.	Various soil dwelling insects.
Cicindelidae (tiger beetles)	Larvae and adults.	Various insects on the ground surface.
Neuroptera		
Lacewings	Larvae and adults.	Aphids.
Diptera		
Syrphidae (hover flies)	Larvae only.	Aphids.
Asilidae (robber flies)	Adults and also larvae in some cases.	Various insects.
Hymenoptera		
Many species of wasps and ants eg, German wasp.	Adults; but prey fed to larvae.	Various insects but especially adult Diptera in the case of German wasp.
Paper wasps		Predominantly caterpillars.
Hemiptera		
Predatory bugs from several families	Nymphs and adults.	Various soft bodied insects.
Arachnids		
Araneae		
Spiders	Juveniles and adults.	Various insects; mostly flying adults.
Acari		
Predatory mites	Juveniles and adults.	Plant feeding mites.
Vertebrates		
Birds, eg, starling		Various soil inhabiting insects.

(b) *Species and stages of hosts attacked*

Most parasitic insects attack only a limited range of hosts and some in fact are confined to a single host species. More commonly however, a group of closely related insects is acceptable. In applied biological control it is often advantageous to have a parasite that will confine its attention to the pest species to be controlled.

All parasitic insects are specialised as to the stage in the life cycle of the host that they attack. Hence some are exclusively larval parasites while others are confined to the pupal or egg stage. Adult insects are rarely parasitised. Eggs may sometimes be laid into one stage in the life cycle of the host but development may not be completed until a later stage. Some parasites thus lay their eggs into the eggs of the host insect but these do not hatch until the host is in the larval stage. Most parasite larvae feed and develop internally within the body of the host insect but the larvae of some species are external and feed on fluids which exude from wounds on the body of the host. One or many larvae may develop within each host individual according to the species of parasite.

Parasites themselves are not immune from parasite attack. They may be affected by other species which are referred to as **secondary parasites** (hyper-parasites). These in turn may be attacked by **tertiary parasites**. Such are the inbuilt checks and balances of nature.

(c) *Life cycle of a typical parasite*

The life cycle of a typical insect parasite is shown diagramma-

Table 14. The main families of parasitic insects and their principal hosts

	Family	*Principal hosts*
HYMENOPTERA		
	Ichneumonidae	Larvae of holometabolous insects, especially Lepidoptera and Hymenoptera (in this case as hyper-parasites).
	Braconidae	Larvae of holometabolous insects, especially Lepidoptera and Diptera. Also aphids.
	Encyrtidae	Larvae and pupae of Lepidoptera.
	Eulophidae	Scale insects, mealy bugs.
	Aphelinidae	Aphids, scale insects.
	Pteromalidae	Larvae and pupae of Lepidoptera, Coleoptera.
	Trichogrammatidae	Insect eggs of various orders.
DIPTERA	Tachinidae	Larvae of Lepidoptera, Coleoptera. Some Hemiptera.

tically in Fig 23. The adults are entirely free living. They may imbibe nectar but do not normally feed on the host insect in any way. Both sexes usually occur but parthenogenesis is common and males may be few in number or absent. Eggs are laid by the female into the appropriate stage of a suitable host insect, usually by insertion of the ovipositor and deposition of eggs into the body cavity. Sometimes however, eggs are simply stuck to the outside of the host or even just deposited nearby. In the latter case the tiny larvae on hatching must make their way to the host insect. This habit is particularly common with some parasitic flies. Once established inside (or externally on) the host the parasite larvae feed and develop. "Non-essential" parts of the host (such as fat body) are consumed first so that the host insect continues to feed and grow. Eventually however, as the parasite larvae near maturity, virtually the entire body contents of the attacked insect are consumed so that little more than the exoskeleton remains. Once mature, larvae of many parasite species (especially Ichneumonoids) emerge and spin a silken cocoon alongside the now dead host. Where many parasite larvae attack one individual a mass of small cocoons is produced. Other parasite species however (especially Chalcidoids) pupate within the dead body of the host insect. To ensure that adults of the parasite are synchronised with the appropriate stage of the host species, emergence from the pupal stage may be considerably delayed, even over the winter in some cases. There may be one or several generations of a parasite each season depending on species.

Pathogens

Insects suffer from diseases caused by micro-organisms in a similar way to higher animals, and such diseases often induce high natural mortality in insect populations. The causal agents are referred to as **pathogens**. Those that produce disease conditions in insects come from the same groups of organisms that are responsible for many diseases of higher animals including man, and thus include bacteria, fungi, viruses, and protozoa. However, the organisms concerned are in almost all cases quite different species so that there is very little if any chance of cross infection. Besides the micro-organisms referred to above, nematodes which attack insects are also sometimes important. These are not true micro-organisms as they are much larger, but they often behave very much like micro-organisms and so are usually considered with them.

Diseases of insects have been intensively studied in recent years and this has given rise to a separate branch of applied science called **insect pathology**. Besides the intrinsic interest in trying to understand the role that diseases play in the natural regulation of insect populations, study of insect pathology has enabled good progress to

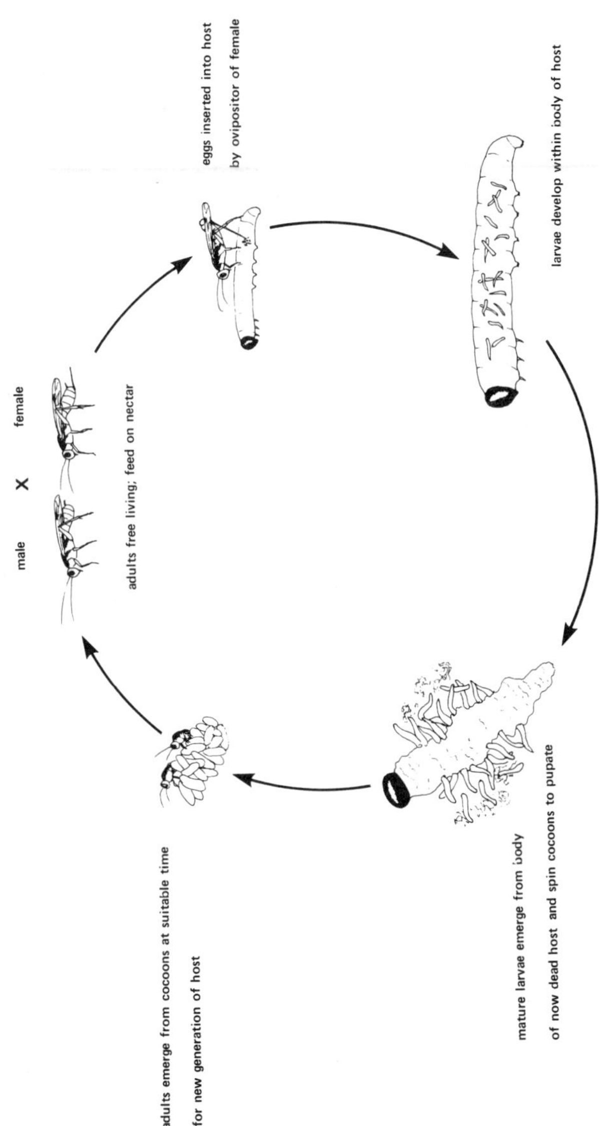

Fig. 23. The life cycle of a typical insect parasite.

be made towards manipulation of micro-organisms for control of pest species. Consideration of the main features of insect diseases and their causal agents is given here and their application in biological control discussed in Chapter 11.

The study of insect diseases has also been important with regard to "domesticated" insects which man has cultured over the ages, in particular the silkworm and the honeybee. In fact the practical importance of insect disease was first realised in connection with maintaining colonies of these two insects. In such situations the disease itself becomes a "pest" to be controlled rather than the reverse.

(a) *Proof of pathogenicity*

The mere association of a micro-organism with a dead or dying insect is not sufficient evidence that it is the cause of the condition. It could be simply an invader of an insect that had been weakened or killed by adverse environmental conditions. To prove that the micro-organism is actually the cause of the disease requires that the well-known **Koch's postulates** be satisfied. Those with knowledge of plant pathology will be familiar with the parallel situation with respect to plant diseases.

Koch's postulates require that:

1. the suspected causal agent be isolated from insects showing the disease symptoms;
2. the micro-organism be capable of growth in pure (artificial) culture;
3. inoculation of healthy individuals consistently reproduces symptoms typical of the disease; and
4. the micro-organism be re-isolated from these inoculated insects.

In practice with insect diseases it is not always possible to follow this procedure completely because many insect pathogens are very difficult or impossible to culture in artificial media with the present state of knowledge. In such cases an insect pathologist may have to be satisfied with the short cut of isolating the organism and using it directly to inoculate healthy individuals with subsequent expression of disease symptoms. However, there is always the danger with this procedure that two similar organisms may occur together and not be separated, with consequent possible confusion.

(b) *The main groups of pathogenic micro-organisms*

Those groups of micro-organisms which include important insect pathogens are listed in Table 15. They comprise the bacteria, fungi, viruses, protozoa and nematodes. Organisms from other groups do occur as disease agents but they are not important. A list of infectious disease agents in man would cover the same groups but their relative importance is rather different with insects. For instance, diseases caused by fungi are relatively much more common in insects than they are in vertebrates.

Also listed in Table 15 are specific examples of insects and their diseases from each group with a brief description of external symptoms in each case. It should be noted that diseases of insects, as with higher animals, do not always rapidly kill the affected individual, if at all. For instance, larvae of Japanese beetle infected with milky disease may progress through all three instars before they finally succumb. In the case of the nematode which affects *Sirex* (steel blue wood wasp), there is no mortality at all but adult female wasps are effectively sterilised so that suppression of the following generation occurs rather than the one directly infected. However, with many other insect diseases high and rapid mortality takes place. This is particularly true for insect diseases caused by viruses.

(c) *Routes of entry and methods of dispersal*

Insects for the most part become exposed to pathogenic micro-organisms either through contamination of their external body surface, or by ingestion of food which contains the organism. There is also the possibility that they may inherit infection from their parents. Four routes of entry are in fact usually recognised:

Table 15.
The main groups of insect pathogenic micro-organisms and specific examples of each group

Group	*Example*	*Symptoms*
bacteria	Milky disease (*Bacillus popilliae*) of Japanese beetle.*	Larvae develop a milky white appearance, become sluggish and eventually die, but action is slow.
fungi	*Entomophthora* spp. of aphids.	Aphids become stuck to leaf surface and surrounded by a white halo of fungal spores.
viruses	Granulosis virus of white butterfly.	Larvae become water-soaked and blackened in appearance. After death the body disintegrates releasing fluid loaded with virus particles.
protozoa	*Nosema* spp. of grass grub.†	Growth of larvae is retarded. Body later becomes abnormally grey and flecked with dark spots due to pathogen development in the fat body.
nematodes	*Deladenus siricidicola* of steel blue wood wasp.	Adult female wasps are sterilised by concentrations of nematodes in the developing ovaries. Males are also infected but are not sterilised.

* — not established in Great Britain.
† — New Zealand species of Scarabaeid.

(1) *Oral* (the micro-organism is ingested). This is the usual route for most viruses, bacteria and protozoa.
(2) *Through the intact integument* (including penetration via the spiracles and tracheal system). This is common for fungi.
(3) *Through wounds* (both external and internal within the gut walls). Some pathogens appear to be able to gain entry only in this way. They may nevertheless be important since wounding of insects both externally and internally appears to be very common.
(4) *Transovarial* (transmission through the egg from the mother). Some viruses may be passed on in this way.

Insects that have been killed by micro-organisms disintegrate sooner or later and release the pathogen into the environment where it is dispersed by wind, rain or the movement of animals such as birds, and other insects. Ability to survive outside of the host may limit the use of insect pathogens as biological control tools.

(d) *Environmental conditions for infection*

For most insect pathogens the presence of infective material in contact with the insect is not sufficient by itself to ensure the development of disease. Correct environmental conditions are also essential. High relative humidity is particularly important for most fungi. The critical importance of environmental conditions for development of infection explains why many insect diseases are sporadic in occurrence and why some attempts to use them for control of pest species have not been successful.

(e) *Specificity*

Most insect pathogens are rather specific in that they affect only a group of closely related insects or even a single species. Some however are considerably broader in activity. Very few are likely to present any significant hazard to humans or other higher animals if utilised for control purposes, but agreement as to what is adequate testing in this respect has not yet been reached.

An individual insect may change markedly in susceptibility to infection with development and early instar larvae for example are commonly much more susceptible than later instars. Adult insects are usually relatively resistant but may become contaminated (by bacteria and viruses in particular) and then transmit infection to their offspring.

SELECTED REFERENCES

Askew, R. R. 1971. *Parasitic insects*. Heinemann, London. 316 pp.
Clausen, C. P. 1940. *Entomophagous insects*. Hafner Publishing Co., New York. 688 pp.
Steinhaus, E. A. 1949. *Principles of insect pathology*. McGraw Hill, New York. 757 pp.
Weiser, J. 1977. *An atlas of insect diseases*, 2nd edition. Junk, The Hague. 240 pp.

Chapter 10

The Ecological Background to Pest Control

We have seen in Chapter 8 that a pest species is only a problem when its population density exceeds the damage threshold level. For different pests and different situations the population that is considered harmful will vary greatly, but the principle remains the same. A consideration of the factors that regulate insect numbers is thus of fundamental importance in relation to pest control. The discussion on these matters that follows is in fact pertinent to *all* living organisms, not just to those we designate as pests, because all species of animals (and plants) are subject to the same basic biological laws. Pests are only a special case because their activities conflict with human interests.

Put in its simplest terms, the numbers of an organism at any particular point in time are a consequence of the interaction between birth rate and death rate. The **reproductive potential** (biotic potential) of a species is always towards an increase in numbers which is counterbalanced by various factors in the environment which tend to restrict or depress numbers. The outcome is reflected in the population density at any particular time. Environmental factors do not remain constant however, but tend to fluctuate in either a regular (eg seasonal) or erratic (eg daily weather) fashion. Insect populations do not therefore remain constant but also fluctuate with time. However, such fluctuations tend to be about a mean which does remain fairly constant over long periods of time, provided drastic alteration of the environment does not take place. In other words, populations of insects (and of other organisms) in the long term must operate on a "replacement only" basis otherwise they would tend towards either extinction or massive increase. The study of changes in insect numbers and of the factors that are responsible for them is that of insect **population dynamics**. It is a fast developing field which to some extent must involve mathematical treatment. Only the briefest discussion of basic concepts is attempted here.

The reproductive potential of insects

A good starting point is to consider the potential rate of increase in numbers of an insect. For most species this is very high and an egg laying capacity of 100 eggs per female is by no means exceptional. Given a sex ratio of 1:1 and normal sexual reproduction, this means a potential population increase of x 50 from one generation to the next (the figure is x 50 rather than x 100 as half the offspring will be males and they do not contribute directly to the next generation). However, in most situations for most of the time various factors prevent this potential increase from actually being achieved. These population limiting factors may operate either at the level of lowering the number of eggs laid (for instance by poor quality adult nutrition), or by mortality of the offspring before they reach adulthood. In natural situations several such factors operate simultaneously. The overall result (of limiting reproductive capacity) is what is commonly referred to as the "balance of nature".

We can depict the process, using our example of a female insect producing 100 eggs, by a simple equation:

Generation 1		*Generation 2*
1 ♂ = ♀	→ 100 eggs →	1 ♂ and 1 ♀
(2adults)		(2 adults)
	98% mortality	

If 100 eggs are produced and the population is to remain constant from one generation to the next, 98% of those eggs, or subsequent stages that hatch from them, must be eliminated before they reach adulthood and become reproductively active. This may seem a very high mortality rate but it is in fact quite normal in nature for any creature that has a high reproductive capacity. Insects do not represent the extreme by any means; some fish for instance produce a million or more eggs and thus the percentage mortality must be correspondingly higher. Spores produced by fungi are more numerous still, but very few survive. It will be clear therefore that what determines the population density of an insect at any particular point in time is just as much a matter of environmental conditions as of the insect's innate reproductive capacity. As environmental factors are so important they are considered in more detail in the following section.

Environmental factors restricting increase of insect populations

We can divide the environmental factors concerned into two groups; those that are part of the non-living environment (physical), and those that are associated with other living organisms (biological). The main factors in each group are listed in Table 16.

By far the most important factor in the environment is climate, particularly temperature because insects are so temperature dependent, though relative humidity may also be important for some insects. Climate sets the limits within which a particular species can exist because of its specific requirements, but in addition the daily fluctuations in climate, which constitute the weather, may be a direct cause of mortality. For example insects may be washed away by sudden rainstorms or killed by frost. Indirectly, climate also has a major influence on the distribution and seasonal occurrence of food plants which may be essential to an insect species' existence.

The second major physical factor is sheer living space. In absolute terms this may not become critical until extreme overcrowding occurs but often there is a special requirement that allows only much lower population densities. For instance, mature codling moth larvae seek suitable overwintering sites in cracks and crevices, or under loose bark, on the trunks and limbs of fruit trees where they spin a cocoon ready for pupation in the spring. If such refuges are inadequate many larvae perish, principally through predation by birds.

Table 16. Environmental factors restricting increase of insect populations

Category	*Factor*	*Density dependence**
PHYSICAL (ie, non-living)	Climate, especially temperature.	I
	Living space.	D
	Soil type (for soil inhabiting insects).	I
BIOLOGICAL (ie, associated with living organisms)	Quantity of food.	D
	Quality of food.	I
	Predators and parasites.	D
	Disease.	D

* — I = density independent.
— D = density dependent. For full explanation see text.

Thirdly, in the case of soil inhabiting insects, various physical features of the soil, such as organic matter content and pH, may be important for some species.

The biological factors fall into two clear groups, those concerned with food supply, and those involving natural enemies whether they be predators, parasites or disease organisms. The nature of these latter factors has been considered in some detail in Chapter 9.

Besides the division of environmental factors into physical and biological groups, such factors may also be classified as to whether they operate in a density dependent or density independent fashion. This requires some explanation. If a factor exerts a similar influence regardless of the density of the population concerned it is said to be density independent. Thus temperature has a similar effect whether there is one insect involved or a thousand. Disease organisms on the other hand often exert much more effect in dense populations than in sparse ones as they can spread much more readily. Such factors are said to be density dependent. The third column of Table 16 indicates whether each factor operates in a density dependent or independent fashion.

Most ecologists agree that density dependent regulators are essential in conferring stability on biological system since they exert correspondingly greater effect as an organism's numbers increase and thus tend to return the population to a lower level. To use a modern expression, they operate on a negative feed-back basis.

The ways that density dependent and density independent factors operate are depicted graphically in Fig 24. The upper line (A) shows the maximum possible rate of increase with all factors at optimum level. This is rarely achieved in nature. With a slightly adverse density independent factor operating the rate of increase is at a lower level but the graph (B) is still an ascending straight line (on a

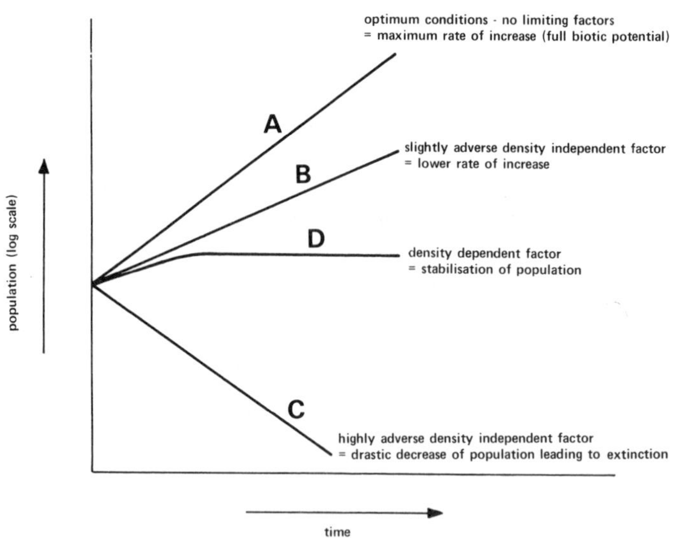

Fig. 24. The effects of density dependent and density independent factors on an insect population.

logarithmic scale), indicating a steadily increasing population. A highly adverse density independent factor produces a marked decline in the population which if unchecked results in extinction (C). Finally, with a density dependent factor the population increases up to a certain point, but as the factor exerts more and more effect, numbers tend to become stabilised and the graph flattens out (D).

In nature several density dependent and independent factors normally operate simultaneously and moreover change with time. It is therefore extremely difficult to sort out the exact reasons for changes in insect numbers over periods of time, and our knowledge in this respect for most pests is very incomplete. Nevertheless the basic principles are firmly established and for some pests good progress is being made in filling out the details. A further important fact to note is that for maximum rate of increase of an organism, *all* factors must be at an optimum. If only one is adverse this is sufficient to limit or prevent population increase.

One final point that may be stressed with respect to regulation of insect populations is that density dependent factors are usually living organisms which respond to changes in the insect population density whereas non-living factors are rarely able to operate in this way. Insecticides therefore, which operate as a density independent mortality factor, can never provide more than temporary control of a pest no matter how effective they may be initially.

Table 17. The structure of a typical insect population

Stage	*Numbers in each group*	*Percentage pre-adult mortality*
eggs	100	40
1st instar larvae	60	20
2nd instar larvae	40	20
3rd instar larvae	20	10
pupae	10	8
adults	2	0
	Total generation mortality	98%

The structure of insect populations

One consequence of the high birth rate/high death rate pattern of most insects is that populations tend to show a preponderance of juvenile forms with progressively fewer individuals as development proceeds towards maturity and additional mortalities occur. We can therefore expand our example of an insect producing 100 eggs in a steady state population to show progressive mortality at different life stages, as set out in Table 17. The figures are given purely as an example and are not meant to denote the exact situation for any particular species or circumstance. The point is clear, however, that inevitably there will be a preponderance of immature stages present for most insects most of the time. Furthermore, it is probably normal for some mortality to occur at virtually every stage in the life cycle as indicated in the Table.

One further, very important point may be drawn from Table 17. If the overall generation mortality were to be reduced very slightly for some reason, from the 98% depicted to say 96%, the consequence at first glance might not be considered significant, as it is a reduction of only 2%, but it would in fact lead to a *doubling* of the population from one generation to the next. Those factors which cause slight but variable mortality may therefore be more important in determining patterns of population change than those which are responsible for high but fixed mortality.

Finally, it might be commented that even with organisms whose rate of reproduction is quite slow, such as ourselves, reductions in mortality can produce dramatic increases in populations in a comparatively short space of time. This is the reason for the explosion of the human population over the last one hundred or so years. The solution in this case of course (in contrast to most pest control situations) is to lower the birth rate rather than to restore mortality to its former level.

The nature and origin of pest problems

Pest species have already been defined as those whose activities are harmful to man. If we consider these more closely, it will be realised that in almost every case the pest is harmful because it is competing with ourselves for some resource. In the great majority of instances the pest wants the resource for food, and although this is often true of humans also, many natural products are utilised for clothing, building materials, decorative purposes and in many other minor ways. The fact remains nevertheless that competition between the pest and ourselves is the basic issue. However, competition between living organisms is a fundamental feature of nature and so the question must be asked as to whether the situation is any different between man and those species he designates as pests. In prehistoric times, when man lived much as any other species of vertebrate animal in a relatively undisturbed natural environment, the answer to this question was probably that there was little or no difference. For modern man however, the situation has changed drastically in several ways:

(a) *Agriculture is "interfering with nature"*

Agriculture by definition inevitably involves some change to the natural environment. This may be slight and temporary, as in shifting peasant agriculture, or drastic and permanent as with modern arable farming. In all cases the natural plant cover is removed (or at least modified) and replaced by a single or a few species of cultivated plants. All competing plants are regarded as weeds and are suppressed as far as possible. Annual cropping, with its cycle of growth, harvest, cultivation and fallow, introduces drastic seasonal changes as well. It is inevitable that these activities will favour some insect species but be harmful to others. If those that are favoured have potential as pests then serious problems can arise.

(b) *Natural communities are complex; agricultural communities are simple*

Most natural communities of plants and animals (ecosystems) are composed of a large number of species which interact in a complex fashion. It is an axiom of ecology that complex communities tend to be stable because they incorporate many checks and balances, and therefore no one species is likely to "explode" in numbers. Agricultural ecosystems in contrast are relatively simple and hence inherently less stable. In such systems rather wide fluctuations in populations of some organisms are to be expected, particularly in annual cropping situations. Outbreaks of pests are thus more likely to occur in agricultural systems than in natural ecosystems. This is no argument for abandoning agriculture though, because there is no other way to feed the burgeoning world population than by intensive agricultural practices. The extreme of simplification in agricultural plant communities is where the same crop (and sometimes even the same variety) is grown over large areas of land in successive seasons (**monoculture**). This is becoming more and more common because of economies of scale and because such previous problems as maintenance of soil fertility can now be managed. Unfortunately monoculture can only lead to accentuation of some pest problems.

(c) *Cultivars have been produced for features other than pest resistance*

The cultivated varieties (**cultivars**) of most agricultural and horticultural crops today bear little resemblance to their original wild ancestors and for many of them we can no longer even be certain of their origins. Over the course of many hundreds of years crop plants have been selected and improved by man for the desired features of higher yield and quality of produce. In this process possible resistance to pests has often been ignored or considered to be of secondary importance. The same can be said of plant diseases but perhaps to a lesser extent. There is little doubt therefore that many present day cultivated varieties of plants are much more susceptible to pests than were their wild ancestors. There are exceptions of course, where more attention has been paid in plant breeding programmes to the development of pest (or disease) resistance, and this has become more common in recent years as its value has been realised. Further discussion of plant resistance as a means of pest control is included in Chapter 11.

(d) *Dispersal of pests by human agency*

The geographical distribution of many living organisms is limited not only by the suitability or otherwise of environmental conditions but also by the fact that they have been unable to reach some parts of the world due to barriers presented by oceans, mountain ranges or climatic zones. This applies particularly to insects but also to many higher animals, to plants, and even to some disease organisms. Man's travel and trade around the world have changed this situation drastically and have provided the means for distribution of many species that would not otherwise occur. Although many plants, vertebrate animals and beneficial insects have been deliberately introduced into many parts of the world, dispersal of many smaller organisms has been fortuitous because they have been able to hitch a ride on man's transport or goods. Many species accidentally introduced in this way have become serious pests because they are largely free of natural enemies in the new area.

Despite the importance of artificial distribution of pest species around the world and the contribution it has made to the pest situation in many places, world distribution of many pests is not by any means complete. For every country there still exists a long list of pest species that have not yet arrived. This is particularly true of geographically isolated countries such as New Zealand. It is the aim of plant and animal quarantine to prevent further unwanted introductions. This is discussed further in Chapter 11.

Some possible solutions to pest problems

It should be apparent from the foregoing discussion that to a large extent pest problems are man made. Although man in his primitive state was no doubt troubled by pests to some extent, the problems would have been much less severe for the reasons stated. If these are the major reasons why pest problems now occur, or at least are more intense, then the possibility must exist of finding solutions by paying attention to the causes. In some cases, such as deliberate breeding of plants for pest resistance, this may be possible. With other causative factors however, solutions may be less obvious or impossible of attainment. For instance, original complex natural ecosystems cannot be re-created once destroyed and would not in any case be agriculturally productive. Total re-creation of the original conditions may not in any case be necessary and manipulation of some environmental features only may suffice. Similarly, although new pest species once introduced are usually impossible to eradicate, control may sometimes be achieved by importation of some of their natural enemies that did not come in with them. These and other more artificial means of dealing with pests are the substance of the next two Chapters.

SELECTED REFERENCES

Cherrett, J. M. Sagar, G. R. (editors) 1977, *Origins of pest, disease and weed problems*. 18th Symposium of the British Ecological Society. Blackwell Scientific Publications, Oxford, 413 pp.

Clark, L. R.; Geier, P. W.; Hughes, R. D.; Morris, R. F. 1974. *The ecology of insect populations in theory and practice.* Methuen & Co. and Science Paperbacks, London. 232 pp.

Southwood, T. R. E. (editor), 1968. *Insect abundance*. Symposium No. 4 Royal Entomological Society of London. Blackwell Scientific Publications, Oxford, 160 pp.

Varley, G. C.; Gradwell, G. R.; Hassell, M. P. 1973. *Insect population ecology*. Blackwell Scientific Publications, Oxford. 212 pp.

Chapter 11

Pest Control Principles and Practices

INTRODUCTION

In this Chapter the main procedures available for dealing with pest problems are discussed but first it is important to be clear just what we mean by the term "control" as applied to pests. To most people effective control of a pest probably means its total elimination from an area, but complete eradication is rarely attainable so that such definition is inadequate. In order to formulate a more appropriate definition we must refer back to the damage threshold concept introduced in Chapter 8. The damage threshold for a pest was defined as the population density below which detectable reduction of yield or quality of a crop does not occur. Effective pest control may therefore be described as reduction or maintenance of a pest population below the damage threshold. It should be appreciated however that decisions whether or not to apply control measures are usually required *before* pest populations reach the damage threshold, and so may be determined on the basis of some lower population referred to as the **economic threshold** (see Chapter 12).

Although the damage threshold concept has very broad application in pest control there are some situations where it is not relevant. For example plant and animal quarantine aims to prevent the spread of pests (and diseases) into areas where they do not occur so that the concept of a threshold level has no validity. A further example is provided by those rare occasions when complete elimination of a pest may be a practical possibility because it has only recently been introduced and is thus of limited distribution.

Categories of crop pests

Individual species of cultivated plants are almost invariably affected by a range of pests of differing importance. Three categories of plant pests may be recognised.

(a) *Key pests*

These are the pests of major importance on specific crops. If left uncontrolled population levels are above the damage threshold most if not all of the time. Effective control of key pests is therefore essential for economic production of the crop concerned. An example is provided by codling moth which is a serious pest almost everywhere that apples are grown.

(b) *Occasional pests*

As the name suggests, pests in this category occur only occasionally in damaging numbers (above the damage threshold). For long periods of time populations remain low and insignificant but every so often, on either a regular or sporadic basis they increase to damaging levels and cause a problem. Leatherjackets are pests of this type. The reasons for fluctuations in populations of occasional pests are often climatic, or due to biological factors such as the incidence of natural enemies, but in many cases the details are as yet unknown.

(c) *Potential pests*

Pests that fall within this category have the potential to reach pest status but are normally suppressed by natural regulating factors. Potential pests are usually recognised only when their natural control is interfered with and they become elevated to actual pest status. Fruit tree red spider mite follows this pattern. Earlier this century it was practically unknown in apple orchards and never reached damaging levels. However, in recent years fruit tree red spider mite has become a major pest in many parts of the world because its natural enemies have been seriously depleted by the widespread use of broad spectrum synthetic insecticides.

The pattern of population change over time with respect to damage threshold levels for these three categories of pests are shown diagrammatically in Fig 25.

PEST CONTROL PROCEDURES — THE MAIN OPTIONS

Most attempts to control pests down the ages have been both primitive and largely ineffective. For crop pests they have included

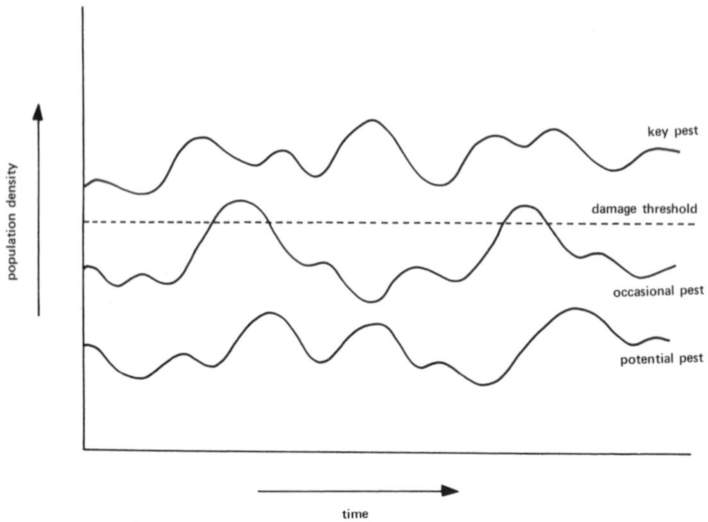

Fig. 25. Categories of crop pests and their population levels with respect to damage thresholds.

such simple measures as hand collecting, beating and trapping. Very limited use was made of naturally occurring poisonous substances.* Until recent times the depredations of pests were thus for the most part accepted as part of the normal scene because so little could be done about them.

During the present century, and more particularly during the past 40 years, vast improvements have been made in pest control techniques, and expectations of pest control by farmers, growers and the general public are now of a much higher order. Although modern pest control has become much more scientifically based, it is not always appreciated that there is still a strong element of trial and error in the development of control procedures. This is because we still have rather inadequate understanding of the complex biological systems within which most pest problems occur and are thus unable to predict accurately the outcome of particular actions.

Pest control procedures fall rather naturally under a number of headings such as chemical control, biological control and cultural control, which are descriptive of the techniques involved and which have until recently been distinct well defined categories. However, some new procedures cut across these established boundaries and

* For those interested, Ordish (1970) has given a full account of the early history of pest control.

furthermore there is now a strong tendency to use two or more approaches together in systems of **integrated control** or **pest management**. (For full discussion of these terms and what they involve see Chapter 12.) Nevertheless, conventional pest control techniques provide (and will continue to provide into the foreseeable future) the main tools of most pest control programmes and the basis for systems of pest management. It is important therefore to be thoroughly familiar with the characteristics of the main pest control options before they are considered for use alone or as components of pest management programmes. The main pest control procedures are listed in Table 18 with an explanation as to what each involves. They are discussed in the remainder of this Chapter.

CULTURAL CONTROL

Cultural control may be defined as manipulation of cultural practices to the disadvantage of pests. In considering the various possibilities that this encompasses it is important to realise that the cultural procedures (and environmental conditions) under which a crop is grown set the limits within which pests can develop. Any change to normal cultural practice, made for whatever reason, may make conditions less favourable for some pests but possibly more favourable for others depending on their individual biological requirements. Deliberate modification of cultural practices to assist with pest control therefore requires a great deal of ecological information about pests if it is to be effective.

In the past, when so few control measures were available, attention to cultural practices to minimise pest problems was important, but with the widespread use of insecticides the possibilities of cultural control were largely neglected. However, total reliance on insecticides for control of pests has given rise to a number of problems, as discussed later, and there is now considerable revival of interest in the potential of cultural control procedures. Unfortunately much of the ecological knowledge necessary for their development and refinement is lacking. Therefore, in the discussion that follows, detailed examples can be provided for only some of the possibilities that come under the heading of cultural control. As with other methods, cultural measures need not provide complete control of a pest. If partial amelioration of a problem is achieved this may be valuable in association with other measures.

The main possibilities for control of pests by cultural means may be considered under a number of headings:

Table 18. The main types of pest control procedures and what they involve

Procedure	*Components*
cultural control	Manipulation of cultural practices to the pest's disadvantage by such means as method and time of cultivation, modification of sowing dates and manipulation of irrigation practices.
plant resistance	The use of species or varities of plants that can grow and produce despite the presence of the pest.
biological control	In the narrow sense involves the use of predators and parasites (mostly other insects) to control pest species. In the broader sense also includes disease organisms (pathogens). Use of the latter is sometimes referred to as microbial control.
chemical control	Principally involves the use of chemicals which are toxic to insects (insecticides) but also includes the use of chemicals which modify insect behaviour, (eg attractants and repellents).
plant and animal quarantine	Involves restrictions on the international movement of plant and animal material to minimise further spread of pests (and of diseases).
mechanical control	Includes killing or trapping pests by mechanical means or the use of barriers to prevent pests from gaining access to plants, stored produce or other materials.
physical control	Involves modification of some physical feature of the environment to render it unsuitable to a pest (eg the lowering of temperature of stored grain), or the utilisation of some physical property as an attractant (eg light traps for night flying insects).
pest management (integrated control)	The blending together of two or more of the foregoing procedures into an overall harmonious system of control. Applies particularly to the integration of chemical and biological control (for detailed discussion see Chapter 12).

Note: In addition to the above there are other specialised pest control procedures (such as the sterile male technique) that do not fit any of the described categories. Because of their limited application, they are not listed in the table and are dealt with only briefly in the text.

(a) Crop rotation

Crop rotation involves sequential planting of botanically unrelated crop plants in normally a three to five year cycle. The practice was developed historically for the purpose of maintaining soil fertility and soil structure, and benefits in suppressing pests have been incidental to a large extent. Traditional rotations of crops have evolved suited to particular soils and climates. Crop rotation exerts most effect on pests which have a narrow plant host range and limited mobility. Their narrow host range means that they are unable to transfer from one crop to the next in a rotation and at the same time their limited mobility does not permit them to reach suitable host plants elsewhere (or readily to re-invade subsequent crops). If additionally they are unable to survive for long in the absence of suitable plants (as is the case with most plant feeding insects) their populations will decline to low levels by the time the rotation has gone full cycle.

These criteria apply to certain soil inhabiting pests and particularly to many plant parasitic nematodes. Crop rotation is particularly valuable as a means of controlling nematodes as they are so difficult to deal with in other ways. It should be noted however that cyst forming nematodes require long rotations as they have the ability to survive in soil for several years in the absence of host plants.

Any move away from crop rotation practices will tend to lead to an increase of some pest problems. Unfortunately, due to developments in fertiliser technology, it is no longer necessary to practise crop rotation to maintain soil fertility and thus there is a trend to more repeated cropping of the one plant species. There are also often economic attractions in repeated cropping with a single plant species (monocropping). Therefore in many farming areas regular rotations of crops are no longer being followed, with consequent increase in severity of some pest problems. In general, the benefits of crop rotation in suppressing pests are often brought to light only when it is abandoned.

(b) Cultivation

Soil cultivation can have a number of adverse effects on soil inhabiting pests. Firstly, mechanical injury may be inflicted such that high mortality ensues. Rotovation is more drastic in this respect than ploughing. Direct injury to insects may not in all cases be necessary as disturbance of such structures as pupal cells can have fatal consequences for the inhabitants. A second effect of cultivation is that bringing deeper soil layers to the surface exposes insects to sunlight, desiccation and bird predation which collectively may deplete numbers considerably. One only has to observe birds following a plough to realise how many soil inhabiting insects must be consumed at such times.

If cultivation is to be deliberately used for dealing with a soil pest a number of factors must be considered which can have a bearing on the outcome. One is the time of year when cultivation is undertaken. For pests with an annual life cycle, vulnerable stages such as pupae may be present at only one time of year. If they are to be affected by cultivation it must be timed to coincide with these susceptible stages. Garden chafer for example can be severely affected by ploughing in late autumn or spring when pre-pupae or pupae are present because if the delicate pupal cell is disrupted it cannot be reformed and the insect dies. Early autumn cultivation has less effect as most of the population is then present as active larvae which are not as affected by soil disturbance. A second factor is the depth to which the soil is cultivated. There is little benefit from ploughing if the pupae are deeper in the soil than plough depth. Thirdly, the type of cultivation and number of passes over the ground may be important. A single well-timed ploughing of pasture for example can provide adequate control of garden chafer but other soil pests such as leatherjackets may require repeated tillage for effective suppression. Unfortunately intensive cultivation is extremely costly and may therefore not be economically justifiable.

As with crop rotation, new technological developments are leading to drastic changes in cultivation practices. In particular the discovery of short lived contact herbicides has permitted the development of minimum tillage techniques which involve little disturbance of the soil compared to conventional methods. Because minimum tillage techniques save time, fuel and labour they are likely to be increasingly adopted. Unfortunately minimum tillage leads to an increase in some pest problems of which slugs are the most important. The factors responsible seem to be lack of soil disturbance, short time interval between the old and the new crop, and a more favourable niche in the seed furrow. The introduction of minimum tillage techniques is a classic example of a technical innovation having unintended consequences and highlights the need for changes to normal cultural practices to be thoroughly evaluated before they are widely adopted.

(c) Time of sowing

Plants are generally most vulnerable to pest injury during the short-lived seedling stage (see Chapter 8). This, together with the fact that many pests have distinct seasonal peaks of activity, sometimes makes it possible to minimise pest injury by careful selection of sowing dates so that the presence of seedlings does not coincide with peak insect numbers. In North America for example, before the introduction of wheat varieties resistant to hessian fly (*Mayetiola destructor*), specific district sowing dates were recommended with the objective of ensuring that wheat seedlings were not present when flies were most active. Similarly, early sowing of spring oats (before mid-March in Southern England or mid-April in the North) will avoid problems with frit fly (*Oscinella frit*) as the plants will then reach the resistant four leaf stage before egg laying commences.

When considering change of sowing date many other practical considerations must be taken into account. These include the ability of the plants to mature within the growing season, risk of injury, and continuity of market supply.

(d) Irrigation

Under conditions of moisture stress irrigation helps to minimise the effects of pest injury because plants are kept growing vigorously and thus are able to compensate to some extent for loss of tissue due to pest feeding (see Chapter 8). However, irrigation can be directly harmful to some pests. The possibilities depend to a considerable extent on the type of irrigation procedure. Trickle irrigation offers few prospects of directly affecting pests but flood irrigation can be harmful to some soil inhabiting species.

Overhead sprinkler irrigation suppresses some foliage pests. Potato moth* for example can be as effectively controlled by frequent overhead irrigation of potatoes as by a programme of insecticide applications. The exact mode of action is not understood but it is known that moist conditions strongly deter egg laying. Also, since eggs are laid either on the plant surface or in the soil adjacent to plants, it is likely that many newly hatched larvae are drowned before they have a chance to tunnel into the plant.

In manipulating irrigation practices for pest control care must be taken that other problems which are favoured by moist conditions, such as some plant diseases, are not accentuated.

(e) Cultural control in perennial crops and in pastures

Perennial crops are ecologically quite different from annual crops in that they present a relatively undisturbed continuum over a number of years, in contrast to the drastic annual cycle of cultivation, sowing, growth and total harvesting of annual crops. There are thus fewer possibilities for dealing with pests by cultural means. Time of sowing for example has no relevance except with respect to initial establishment, and soil cultivation may have little application. Other measures however, which are not applicable to annual crops, may be amenable to manipulation. Pruning for example may enable portions of plants infested with pests to be removed and destroyed. Such practices are recommended for dealing with the shoot tunnelling larvae of currant clearwing (*Syanthedon tipuliformis*). Shelter belt trees which are alternative hosts for orchard

* A pest not established in Great Britain.

pests such as leaf rollers can be a serious source of infestation. Their replacement with non-susceptible species aids control of these pests and provides another example of a cultural control measure in a perennial crop situation. An important aspect of perennial crops is that they present a more stable environment than annual crops within which biological control factors for pests can develop and be maintained, but this falls within the sphere of biological rather than cultural control.

At first glance there seems little possibility of applying cultural control practices in pastoral farming. Type of cultivation and time of sowing for example have relevance only at pasture establishment, and rotation is meaningful only with respect to short term pasture (leys). Management of grazing animals, however, is a factor of considerable importance with respect to pest incidence in pastures. It is known for example that continual close grazing of ryegrass swards, especially during the first few years from establishment, can encourage the build up of frit fly by prolonging the susceptible stage of plant regrowth. On the other hand in New Zealand it has been shown that intense grazing at critical periods in the life cycle of certain pests (eg, while eggs and early larval stages of some soil inhabiting pests are on the soil surface) can seriously deplete populations. Detailed knowledge of the pest's life cycles and habits is essential if maximum benefit is to be gained from such practices. Unfortunately adequate information for many pasture pests is not yet available.

Cultural control — some conclusions

Cultural control measures which simply involve modification of some essential cultural procedure (such as change of sowing date) are cheap because the operation has to be carried out anyway. However, if cultural operations have to be specially undertaken (such as cultivation and re-sowing of pasture) they can be very costly.

Cultural methods of pest control present no toxicity or residue problems (in contrast to the use of insecticides) and harmful effects on non-target organisms are minimal. Unfortunately the applicability of cultural methods for some pests is limited and, because cultural control is often preventative in nature, it is usually necessary to plan ahead to prevent a pest problem arising rather than take action after it has developed. Despite its limitations it may be concluded that cultural control has an important role to play in dealing with pests of plants and that attention to cultural practices, as they may influence pest incidence, should be a consideration in any agricultural or horticultural enterprise.

PLANT RESISTANCE

Plant resistance to pests may be described as the ability of a plant species or variety to grow and produce economically despite the presence of the pest. For the most part resistance is concerned with differences between cultivated varieties (cultivars) and hence the term **varietal control** is sometimes used. However, those plant species that fall outside the normal host range of a pest may be regarded as resistant and in some situations it may be possible to utilise such resistant species to deal with a pest problem. The term plant resistance is therefore used in this book in its broadest sense to include both resistant cultivars and resistant species of plants. The latter are discussed first.

(a) Pest resistant plant species

One situation where advantage may be taken of pest resistant plant species is in selection of ornamental plants for landscape development or improvement. Here the exact species of plants may not be important provided they have the appearance and growth characteristics desired by the horticulturalist. In selection of trees for urban street planting for instance, freedom from serious pest attack should be an important consideration because of the difficulty and expense of other means of pest control, such as insecticide spraying.

In some cases there may be a large number of species within a single genus of plants from which to choose and which differ in degree of resistance to a pest. The genus *Eucalyptus* for example includes many species which range from highly susceptible to somewhat resistant to the foliage feeding eucalyptus tortoise beetle (*Paropsis charybdis*).

The natural resistance of some timbers to pest attack is dependent on tree species. The selection of pest resistant timbers for building construction is particularly important in tropical countries where pests such as termites abound and where chemical timber preservation is not widely practised.

Pastoral farming also provides considerable opportunity for the use of pest resistant plant species. In many temperate parts of the world pastures are based largely on ryegrass and white clover because of the growth patterns and nutritional value to stock of these plants. Unfortunately these species are susceptible to a wide range of pests. There is no theoretical reason why white clover and ryegrass should not be replaced by other more pest resistant plants provided these produce animal feed in sufficient quantity, of suitable quality and with acceptable seasonal growth patterns. To some extent lucerne, which is resistant to certain pests (eg, the

Scarabaeid, *Costelytra zealandica* in New Zealand), can be used to replace pasture legumes and grasses. Lucerne has the added advantage that it is drought resistant due to its deep tap root. Unfortunately lucerne is highly susceptible to several species of aphids. This emphasises the point that, in making such changes, care must be taken to ensure that one pest problem is not simply exchanged for another. The full possibilities for development of pest resistant pastures have certainly not yet been explored and further progress in this direction may be expected.

(b) Pest resistant cultivars

That differences in susceptibility to pests (and diseases) do occur between cultivars of many plants is a common observation of horticulturalists and home gardeners. Sometimes a pest problem may be minimised by deliberate selection of resistant cultivars, but more often than not growers choose cultivars for other desirable characteristics such as high yield or quality of produce and worry about pest problems afterwards. Similarly, in plant breeding many cultivars of horticultural and agricultural plants have been developed primarily for features of high yield and desirable market quality rather than for pest resistance. This applies to a lesser extent to disease resistance where more emphasis has been placed on the development of resistant cultivars probably because plant diseases are generally more difficult to control by chemical (or biological) means than are most pests. There are some examples however of deliberate and successful breeding for resistance to pests in crop plants (see below) and attention to this is now becoming more common in plant breeding programmes as its importance and potential are realised and because of problems encountered with other means of control.

One of the earliest examples of the successful development of a pest resistant plant is that of wheat cultivars resistant to hessian fly (*Mayetiola destructor*) in North America. Last century hessian fly was a very serious pest in North American wheat growing areas but has been effectively suppressed for many years by fly resistant cultivars. In this example plant resistance provides complete control of the pest in question but a number of different resistant varieties have had to be produced to cope with the problem of pest bio-types (see page 180).

Another example of long standing successful control of a pest by plant resistance (although not strictly a varietal one) is that of *Phylloxera*, a serious pest of vines. *Phylloxera* almost wiped out the European wine industry about 100 years ago. After many unsuccessful attempts at control by various means a solution was finally found which involved grafting European vines onto rootstocks of American vine species (the insect is primarily root feeding). The American rootstocks were resistant to the pest but could not be used without grafting because of inferior wine quality. Virtually complete control has been provided by this method up to the present day.

The final example of varietal resistance to pests concerns three aphid species which attack lucerne in New Zealand. Only one of these aphids (pea aphid, *Acyrthosiphon pisum*) occurs in Great Britain.

Blue green lucerne aphid (*Acyrthosiphon kondoi*) first appeared in New Zealand in 1975 and rapidly caused serious damage to lucerne throughout the country. Cultivars of lucerne resistant to this pest had already been produced in North America and some of these have been introduced into New Zealand. Breeding for resistance has also been initiated there. The situation is complicated by the fact that a second species of aphid (pea aphid, *Acyrthosiphon pisum*) has more recently been discovered in New Zealand. A third very serious aphid pest of lucerne, the spotted alfalfa aphid (*Therioaphis maculata*) is present in Australia and it is probably only a matter of time before it too gains entry into New Zealand*. The problem therefore may be one of trying to produce a lucerne cultivar resistant to all three aphids rather than to one, and at the same time to retain or improve resistance to other problems such as stem nematode and wilt diseases.

(c) Mechanisms of plant resistance to pests

Most pest resistant cultivars have been developed by exposing plants to the pest over several plant generations and selecting for further breeding those that survive best. Understanding of the underlying mechanisms of resistance is thus not essential but such a state of affairs is not satisfactory from a scientific point of view. Also it is likely that understanding of the mechanism(s) of resistance would lead to more rapid and effective development of resistant cultivars. Considerable effort has therefore been devoted to this problem and the essential features of plant resistance to pests are now established though many details remain to be elucidated.

Three categories of plant resistance are usually recognised:

(1) *Non-preference*

The pest chooses not to feed or oviposit (or feeds less or lays fewer eggs) on the resistant cultivar compared to susceptible ones. Non-preference may only be exhibited when the insect is offered a choice, but in its extreme form it operates whether there is a choice

* — detected in late 1982.

or not; (the insect starves rather than feed on the resistant cultivar). Non-preference is usually due to the presence in resistant plants of chemical substances which deter feeding or oviposition but may also be due to the absence of chemical attractants or stimulants compared to susceptible plants.

(2) *Antibiosis*

In this case the pest feeds, or attempts to feed, on the resistant plant but does not develop properly. Growth of the insect is often slower compared to that on susceptible cultivars. Smaller adults may result or larvae may die before reaching maturity. Egg production also may be seriously curtailed. Antibiosis is thus directly harmful to insects and is due primarily to the presence of toxic substances in the plant though lack of essential nutrients may also play a part.

(3) *Tolerance*

Plants exhibiting tolerance, because of their growth characteristics and vigour, are able to grow and produce despite insect injury. Feeding deterrent or toxic chemicals are not involved and the pest is unaffected by feeding on tolerant plants.

With individual pest resistant cultivars more than one of the above mechanisms may operate simultaneously. Thus a plant may exhibit some antibiotic properties towards a pest but may at the same time also display vigorous growth and thus be tolerant to some degree of pest injury. In practice therefore distinction between the three categories of plant resistance is not always precise and clear cut.

(d) Pest biotypes

After a pest resistant variety has been grown for some time resistance may appear to break down. In fact this is due to the development of a strain of the pest (**biotype**) which is able to overcome the plant's resistant properties rather than to any change in the plant itself. The process is very similar to the way in which pests become resistant to insecticides (see under Chemical Control later in this Chapter) and takes several generations of the pest for its appearance. Similar breakdown of plant resistance to diseases is a serious problem but fortunately with insects it is much less common. It is most likely to occur with pests of tropical crops such as rice, where several generations of the pest may occur within the life of a crop and several crops may be grown each year so that rapid selection can take place. Resistance is also more likely to be overcome by pests where it is due to a single inherited characteristic of the plant rather than to several factors operating together.

Plant resistance to pests — some conclusions

For the most part in the past insufficient attention has been paid to the possibilities of developing resistant plants as a means of dealing with crop pests. This is perhaps due in part to the fact that complete rather than partial solutions to pest problems have been sought in this way. Although there are some examples of plant resistance providing virtually total control of a pest, partial resistance is much more common. Only in recent years has it been appreciated that partial resistance may be of great practical value in conjunction with other measures. Partial plant resistance for example may enable the number of insecticide applications normally applied to a crop to be reduced by half or more, resulting in considerable cost saving. Similarly, biological control measures (see next section) are more likely to provide an acceptable level of control where tolerant plants are concerned because natural enemies rarely eliminate pest populations completely. The greater use of partially resistant cultivars is thus an important objective in the development of many pest management programmes (see Chapter 12).

Plant resistance is a very cheap form of pest control. It is therefore a particularly attractive approach to pest problems in low value crops which may not generate sufficient profit to justify more costly measures. This applies to most pastoral farming and particularly to peasant agriculture.

The potential for development of pest resistant cultivars of most crop plants has not yet been fully explored by any means. When resistant properties are sought within a crop species, or in closely related plants, it is usually found in some degree. However, resistance initially may be combined with undesirable features of poor quality or low yield and it is then the task of the plant breeder to try and incorporate resistance with more desirable features in new cultivars. This often takes a long time and several years may elapse between initial discovery of resistant plant material and the availability of new cultivars for field evaluation and release. Unless resistance is combined with other desirable qualities it is of little practical value. New pest resistant cultivars must be acceptable to farmers and growers in terms of yield and quality of produce if they are to be adopted.

BIOLOGICAL CONTROL

Biological control may be defined as the use of natural enemies to suppress pest species. The term natural enemy refers principally to parasites and predators (mostly other insects) but may also include disease organisms. Use of the latter is sometimes considered separately under the heading of **microbial control**. Biological control always involves some degree of deliberate manipulation which distinguishes it from **natural control**. The latter refers to the natural *regulation* of insect populations that is going on all the time without human interference (see Chapter 10) and includes non-living as well as living components of the environment.

It has sometimes been suggested that the scope of biological control should be broadened to include such procedures as the sterile male technique, the use of insect pheromones and even plant resistance to pests, on the basis that all such measures involve living organisms or biologically active substances derived from them. Such a wide definition seems undesirable as logically it should include all control procedures that are not chemical or physical in nature. The term biological control is used here in its normal narrower sense and other biologically based control procedures are discussed separately.

An introduction to insect predators, parasites and diseases of insects induced by micro-organisms was given in Chapter 9. This section deals with the more important aspects of their practical utilisation for suppressing pests.

(a) Biological control procedures

There are three distinct types of biological control procedures that can be applied using parasites, predators and disease micro-organisms.

(1) *Inoculation (introduction)*

This involves the introduction of new species of parasites, predators, or disease micro-organisms into areas where they did not previously occur. Once introduced and established they are left to spread and maintain themselves as best they can though dispersal initially may be assisted by artificial means. Introduction of parasites and predators is the classical biological control procedure and is that employed in most examples of successful biological control over the past 100 years. If the pest to be controlled has itself earlier gained entry to the country or area from some other part of the world, the prospects for success are usually quite good as natural enemies are unlikely to have come with it. The biological control project in such cases involves the search for natural enemies in the area of origin of the pest and the introduction of those considered most suitable. Strict quarantine is essential to ensure that other unwanted organisms such as hyperparasites (parasites of parasites) are not also accidentally introduced. Biological control by introduction of new species of parasites, predators and disease organisms is a function of centrally funded research organisations and cannot be undertaken by individual farmers or growers.

Indigenous pests (those that occur naturally in a country) reach pest status because they can thrive under agricultural conditions whereas their natural enemies are unable to do so. Compared to introduced species, indigenous pests offer much poorer prospects for biological control by inoculation procedures because their natural enemies are already present, though ineffective. The only possibility is to locate natural enemies of closely related insects in some other part of the world and introduce these in the hope that they will attack the pest. Very few successes have been recorded with such an approach. However, the permanent benefits from successful biological control are such that inoculation procedures are worth attempting for most serious pest problems even though the prospects for success are considered to be poor.

It is not possible to predict to any extent the effectiveness of parasites or predators before they are introduced. A parasite that appears insignificant in one place may be quite effective when introduced elsewhere and vice versa. The only way to find out what the effect will be is to make the introduction and see what happens. Some people have questioned whether this might not be a risky thing to do as there seems to be no guarantee that the parasite will not attack beneficial organisms. Such however is very rarely the case because insect parasites are highly specific and cannot adapt to "unnatural" hosts. The practical record of biological control in this respect is extremely good. However, problems can occur with vertebrate predators such as birds which are often much more flexible in their feeding behaviour. For this reason the deliberate introduction of such larger animals for biological control purposes is now not generally favoured.

There has been little opportunity for biological control by inoculation procedures in Great Britain because most pests are indigenous and thus not amenable to suppression by introduced parasites or predators. There is also the problem that cool climates with a long dormant season do not provide conditions for many parasites to flourish. The best examples therefore of biological control in Britain come from glasshouse situations where the introduced parasitic wasp *Encarsia formosa* can provide effective control of glasshouse whitefly (*Trialeurodes vaporariorum*) and the predatory mite *Phytoseiulus persimilis* can hold populations of the two-spotted spider mite (*Tetranychus urticae*) in check. However, both species require careful management and seasonal re-introduction and thus their utilisation falls within the scope of pest management (see Chapter 12) rather than biological control in the pure sense.

(2) *Mass rearing and release (augmentation)*

The possibility was suggested many years ago of rearing large numbers of a parasite or predator already present in an area and releasing them into field situations at appropriate times to give short term local control of a pest. The idea was to boost numbers of the parasite or predator before pest populations became too great and to overcome to some extent the delay in increase of natural enemies which always occurs following increase of pest populations. Development of the technique has been limited by the difficulties and large labour requirement involved in rearing most parasites and predators. Major advances in methods of rearing insects in large numbers, including the development of artificial diets, have given renewed impetus to the technique. It is still not practical to rear parasites directly on artificial diets but some predators can now be raised in this way in large numbers at low cost. Some lacewings and ladybirds for example are now being produced commercially in North America for sale to farmers and growers. Parasites have to be reared on suitable living host insects but if these can be produced easily the whole process becomes more of a practical possibility.

Mass rearing and release of parasites and predators as a form of biological control is not yet in widespread use in Western countries and is not at present practised in Great Britain (except for seasonal introduction into glasshouses). However, in some parts of the world (Russia, China) considerable effort is being devoted to it and claims made in particular for the effectiveness of parasites of insect eggs, such as species of the tiny parasitic wasp *Trichogramma*. These insects are extremely small and delicate, and their effective use depends on accurate timing of releases in relation to egg laying of the pest(s) to be controlled, so that considerable expertise is involved. Despite these problems the technique offers considerable potential which is as yet largely unrealised.

Most uses of disease organisms for pest control involve a process of augmentation whereby the micro-organism is cultured artificially and then re-introduced into the environment at a suitable place and time. This topic is discussed further later in this section.

(3) *Conservation and encouragement*

This form of biological control aims to make better use (by various techniques) of natural enemies that already exist in an area. It merges into and is an important part of pest management.

Conservation of natural enemies is concerned for the most part with minimising harmful effects resulting from the use of insecticides. Most insect parasites are much more susceptible to insecticides than the pests they attack and so their populations are very easily depleted following chemical treatment. Such harmful effects can be minimised by careful selection of insecticides and by limiting the dosage and frequency of application. In orchards for example predatory mites and insects, if conserved, can provide effective control of fruit tree red spider mite, though some assistance from selective acaricides may be necessary. This topic is discussed further in Chapter 12.

Encouragement of natural enemies implies something more positive than merely eliminating harmful effects. It is known for example that the adults of many parasitic Hymenoptera (wasps) have a requirement for nectar if they are to reproduce successfully. Provision of suitable flowering plants in the vicinity of crops may meet this need and thus boost the populations of such parasites. Techniques for encouragement of natural enemies are limited at present but as more detailed knowledge of the ecological requirements of parasites and predators is developed there will be greater opportunity for their successful manipulation.

In the long term, conservation and encouragement of natural enemies may prove to be the most important of all biological control procedures because virtually all pests are attacked to some extent by parasites and predators which have the potential to provide improved control if manipulated to best advantage. The same may be said of insect diseases.

(b) Biological control by insect pathogens

Micro-organisms causing disease in insects (insect pathogens) can be regarded as specialised natural enemies. Although pathogenic micro-organisms differ greatly in size and lifestyle from insect parasites and predators the principles of their exploitation for pest control are much the same. Thus we can distinguish the same procedures of inoculation, augmentation and conservation as for parasites and predators. However, with disease micro-organisms augmentation is the dominant technique and inoculation and conservation procedures play comparatively minor roles. Most practical uses of micro-organisms for pest control thus involve their culture in artificial media and later introduction of comparatively large amounts of inoculum into the environment at an appropriate time and place. The technique is possible only with those micro-organisms that can be readily cultured in artificial media and moreover that can be induced to produce spores or other suitable resting stages which permit storage and application. Many (but not all) fungi and bacteria can be handled in this way but insect viruses have the limitation that they have to be raised in living insects. This is unfortunate as insect viruses probably have greater potential for pest control than any other group of micro-organisms because of their virulence and selectivity.

A few insect pathogens that can be cultured without too much difficulty and which produce a suitable resting stage for storage and dispersal have been developed as commercial products. The best known example is the bacterium *Bacillus thuringiensis* which is effective mainly against larvae of Lepidoptera and Diptera. The killing effect of *Bacillus thuringiensis* is due primarily to chemical toxins which it produces. It does not sustain itself to any extent in the field and so has to be used rather like a chemical insecticide. There have been difficulties in manufacturing a standardised product so that the effectiveness of commercial preparations has not always been consistent and the material has somewhat fallen into disrepute. This is unfortunate as, like most insect pathogens, it is selective in action and does not harm parasites or predators of pests to any extent.

Another bacterium, *Bacillus popilliae*, is the cause of milky disease of Japanese beetle (a close relative of the garden chafer) in the United States. *Bacillus popilliae* can be grown in artificial culture but does not readily produce spores. Some commercial production has been undertaken by raising it in living insects but this is very expensive and can only be justified for control of Japanese beetle in high value turf areas such as putting or bowling greens. It is uneconomic for pasture treatment.

Amongst insect pathogenic fungi, commercial preparations of *Verticillium lecanii* are now available (in three forms) for control of aphids, whitefly and thrips under glass. Effectiveness out of doors is unlikely to be reliable because of the requirement of high humidity for infection to take place.

No viruses are at present used in Great Britain for insect control but research has been undertaken with pathogenic viruses of codling moth and other pests.

Simple introduction (inoculation) of new insect pathogens into areas where they did not previously occur is a less important technique than it is with parasites and predators for biological control and there are as yet no examples from Great Britain of widespread control of a pest by such means, but many insect pathogens have undoubtedly been inadvertently spread around the world, as have many pests.

The natural occurrence of diseases caused by micro-organisms is common in insect populations and is a major natural mortality factor in many situations. However, little progress has been made in exploiting such naturally occurring infections for pest control and in practice techniques of encouragement aimed at enhancing the effect of insect diseases are unimportant. One reason for this is that disease development is usually dependent on optimum combinations of environmental conditions (of temperature and humidity in particular) and there is little opportunity to regulate these in crop environments. Another difficulty is that high population densities of host insects may be necessary for disease outbreaks to develop. Progress has also been limited by current lack of knowledge of the causal micro-organisms and how they might be influenced by cultural practices.

Biological control — some conclusions

Biological control has an important role to play in modern pest control programmes but can never provide a complete solution to *all* pest problems. Indigenous pests in particular can rarely be controlled by the introduction of new species of parasites or predators. A further difficulty is that the degree of control provided by natural enemies is rarely adequate for pests which cause injury directly to the harvested portions of crops. Horticultural produce in particular must be blemish free for market acceptability and the high standards of pest control necessary to achieve this cannot usually be provided by biological means. In other situations however, particularly in indirect agricultural production systems such as pastoral farming, any reduction in pest populations brought about by natural enemies is valuable. A big advantage of biological control is that it lacks many of the problems associated with chemical control procedures. There are for example no harmful effects on other beneficial organisms nor are there hazards to humans or wildlife. Also, pests do not become resistant to natural enemies in the way that they do to insecticides. A further point is that biological control once established is largely self sustaining as natural enemies tend to respond to increases in the populations of their hosts. For these reasons renewed impetus has been given to biological control in recent years, particularly as a component of pest management systems. In such systems biological control is supplemented by other methods such as the carefully controlled use of selective insecticides.

There is little doubt that the potential for biological control of many pests has not yet been fully explored and further advances may be expected both in situations where it provides the sole means of control of a pest and where it is used in conjunction with other measures.

CHEMICAL CONTROL

Chemical control in the conventional sense involves the use of chemicals (insecticides) to kill pests. However, in recent years many discoveries have been made of substances that influence insect behaviour, for example chemicals which are attractive to pests, or which produce some other effect such as sterility, without killing the insect concerned. Where such substances can be exploited to manipulate pests and provide or help to provide solutions to pest problems their use clearly falls within the scope of chemical control. A modern definition of chemical control might therefore be— "The use of chemicals to kill, deter or in other ways influence pests for control purposes".

This section is concerned with a discussion of the basic principles of pest control with insecticides and the major points which need to be considered when using them in practice. Detailed recommendations for chemical control of individual pests do not fall within the scope of this book.

Since World War II insecticides have played a dominant role in pest control and, despite increasing attention to alternative means of dealing with pest problems, continue to do so in most situations. There are a number of reasons for this. Firstly, insecticides are usually rapid and curative in effect so that action can be taken quickly against a pest problem which threatens a grower's crop. Moreover such action can be undertaken by the farmer or grower himself so that he does not have to rely on the services of a specialist. A further point is that application of insecticides can provide a very high level of pest control and thus enables high quality blemish free produce to be grown. In many branches of horticulture in particular, such as fruit growing, top quality produce is essential to ensure good market prices and profitability. Whether such levels of control can be permanently sustained however is questionable, as discussed later.

Inherent in the use of insecticides for pest control are a number of drawbacks, as with other control methods. A major one is that insecticides for the most part are not very selective for the pest to be controlled and to some extent also affect other living organisms. If these are parasites or predators of the pest against which treatment is aimed, control initially may be good but when the pest population recovers (as it inevitably does in time) it may rise to higher levels than before because some of its natural controls have been destroyed, a phenomenon known as **resurgence**. Insecticides may also eliminate natural enemies of potential pests so that they are raised to pest status. A further group of beneficial insects that may be harmed by insecticides are pollinators of cultivated plants, bees in particular.

Another aspect of the rather non-selective action of insecticides is that many are toxic to higher animals resulting in harmful effects on wildlife. There may also be hazards to humans. These take the form of risk of poisoning for those manufacturing and applying pesticides and of traces of chemicals (**residues**) associated with harvested produce.

A point of great practical concern for continued effectiveness of chemical control is the widespread ability of pest species to develop resistant strains. With many crop and public health pests this is a very serious problem for which there is as yet no entirely satisfactory answer.

Finally, although the great majority of uses of chemicals for pest control have been profitable up to now, rapidly increasing costs of manufacture are drastically changing the situation and many farmers and growers are questioning whether such expenditure can continue to be justified. For some crops and some situations the answer may well be that it cannot. In any case, chemical pest control inevitably involves *continual* expenditure in contrast to some other methods, such as biological control by inoculation, where the initial cost may be the only cost.

Most problems associated with chemical control arise from the fact that use of an insecticide against a crop pest does not simply involve action of the chemical on the pest alone. This simple interaction can be represented by the notation:

$$\text{chemical} \rightarrow \text{pest.}$$

In fact the interaction is between the chemical and a complex biological system of which the pest is only one component. This interaction is more truly depicted by:

$$\text{chemical} \rightleftarrows \text{ecosystem in which the pest occurs.}$$

Failure to recognise the complexities that may be involved accounts for many problems in the use of insecticides. The more important undesirable consequences of chemical control of crop pests are summarised diagrammatically in Fig 26.

(a) Origin of chemical control

Although pest control with synthetic insecticides is a modern development the use of chemicals to help limit the ravages of pests dates back many hundreds of years. The substances initially available were naturally occurring, including some derived from plants, or in more recent times by-products of industrial manufacture of some sort. Most were not very effective as insecticides and were never widely used. Furthermore, equipment capable of applying them in a satisfactory manner was not available. Modern insecticide development only really began with the discovery of the potent insecticidal properties of DDT in 1936-37. This was followed during

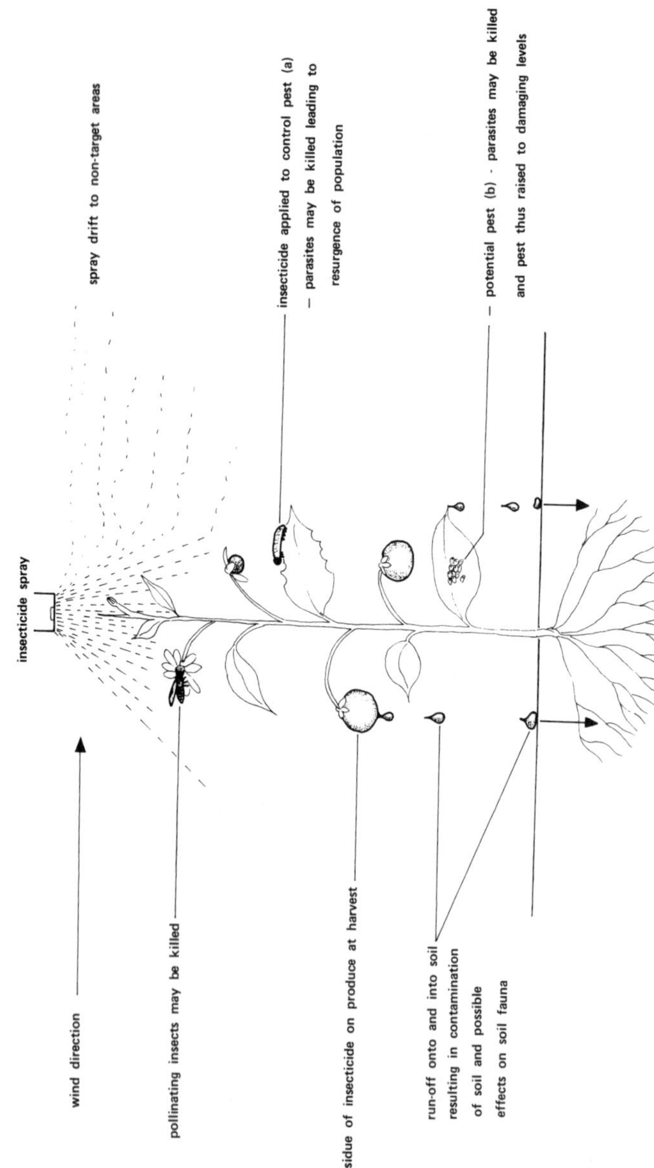

Fig. 26. Some important consequences of the use of insecticides for pest control.

and immediately after the war by a great expansion of chemical synthesis of organophosphorus compounds. Initial discovery of the insecticidal value of the latter group of chemicals arose from war-time research on poison gases. Widespread chemical control of insect pests is therefore of very recent origin, a fact which is not always sufficiently appreciated.

(b) Discovery and development of insecticides

The discovery and development of new pesticides (this term includes fungicides and herbicides as well as insecticides) is nowadays almost entirely within a small number of large commercial companies around the world. The costs of synthesis, evaluation, safety testing and market development are now so great (£10 – 15 million per new product) that only very large organisations can afford such investment. Additionally there is very uncertain return of profits. Initial discovery of new insecticides may sometimes be made in government laboratories, as has been the case with the new synthetic pyrethroids, but practical development is then almost invariably placed in the hands of commercial companies because these are better geared to handle manufacture, promotion and marketing.

Since there is still no way of accurately predicting the biological activity of new chemical compounds the only way to discover new pesticides is by massive chemical synthesis and testing. The ratio of marketable products to new substances synthesised in the laboratory is only about 1 in 10,000, which is one reason why costs of discovery and development are so high. Besides proving the effectiveness of new compounds, the companies concerned also have to undertake extensive toxicity testing of any materials they wish to market. Not only is this another added cost, but such toxicity testing is a lengthy process and results in a time span of at least five years between initial discovery of a potential new insecticide and marketing. During this time no profit at all is earned. It is small wonder therefore that pesticides are expensive to the end user.

(c) Chemistry

There are many different insecticides on the market and if one multiplies these by the variety of trade names the array is bewildering. However, as far as the **active ingredients** (insecticidal chemicals themselves) are concerned they fall into quite a limited number of chemical groups, as follows:

Organochlorines

These, as the alternative name chlorinated hydrocarbons

suggests, consist of a basic hydrocarbon nucleus (straight chain or ring structure) with various numbers of chlorine atoms attached or substituted at places in the molecule. DDT, dieldrin and lindane are well known examples. The organochlorines are the so-called "hard" insecticides because they generally persist for a long time in the environment. This property, once regarded as desirable for long term pest control (and still so in some situations, eg termite protection of timber), gives rise to problems with residues in harvested produce and to undesirable contamination of the environment. Although formerly used extensively for crop pest control, the organochlorine insecticides have now been largely phased out (as for example aldrin for routine treatment of soil to control wireworms), and substituted by less persistent but not necessarily less toxic materials.

Organochlorine insecticides act by contact, ingestion and also in some cases by vapour action. None are systemic*. (See section headed Mode of Action on p. 195 for further definition of these terms.) Toxicity to humans varies greatly according to the particular insecticide concerned and there is no effective antidote for mammalian poisoning (in contrast to organophosphate and carbamate insecticides).

Organophosphates

These all have a similar central core to the molecule which includes a phosphorus atom. The rest of the molecule varies greatly. There are many organophosphate insecticides on the market each differing in persistence, effectiveness against different pests, and toxicity to humans. The group includes some of the most toxic insecticides in use (such as parathion) as well as some of the safest (such as malathion). Consideration of individual chemicals is therefore essential in assessing precautions to be taken in handling and applying them.

All organophosphate insecticides are nerve poisons both for insects and humans, and inhibit the enzyme cholinesterase. Atropine sulphate and PAM (Pralidoxim) are effective antidotes in cases of human poisoning but must never be given unless organophosphate poisoning is definitely known to have taken place as they are otherwise toxic themselves.

Some organophosphate insecticides (eg, demeton-s-methyl) have systemic properties and are particularly effective against sap sucking insects such as aphids. Many organophosphate insecticides also act by contact and some (dichlorvos for example) exert considerable fumigant effect.

* Absorbed into and translocated by the sapstream of plants.

Carbamates

Carbamate insecticides are all based on the carbamic acid nucleus. Like the organophosphates they are cholinesterase inhibitors but far fewer have been developed as insecticides. The best known is carbaryl. Some carbamate insecticides (such as carbaryl) have very low toxicity and can be recommended for home garden use, but others (such as aldicarb) are extremely poisonous and their sale and use are strictly controlled. Atropine sulphate is an antidote for poisoning by carbamate insecticides (as for organophosphates) but PAM should not be given.

Plant derivatives

The fact that certain plants have toxic properties for insects has long been known and plant derived insecticides must have been amongst the earliest used. Until very recent times there has been considerable use of materials such as pyrethrum, rotenone and nicotine (all extracted from plants), but they have now been largely superseded by synthetic products. A popular misconception is that naturally occurring chemicals are necessarily safer than man made products. Some, such as pyrethrum, do have very low toxicity but others are extremely poisonous and nicotine is among the most hazardous of all insecticides. As with synthetic insecticides therefore each chemical must be considered individually and it is impossible to generalise.

Synthetic pyrethroids

Natural pyrethrum has broad insecticidal activity combined with low mammalian toxicity, but it has never proved effective for plant pest control because it breaks down very rapidly in sunlight on plant surfaces. However, the past few years have seen a major breakthrough in the development of effective synthetic pyrethroid insecticides with the culimination of many years patient research at Rothamsted Experimental Station. This has led to the introduction of a whole new class of insecticides (**synthetic pyrethroids**) which are being developed under licence by several major chemical companies. Some of these chemicals (such as deltamethrin and permethrin) are now being marketed for control of various crop pests. The synthetic pyrethroids are generally much longer lasting than natural pyrethrum and have high insecticidal activity against a wide range of pests. They are considerably more toxic to humans than the natural product but still within the range of acceptability for most purposes. They are extremely expensive per kilo, but this is largely compensated by the fact that they can be used at extremely low rates (as little as a few grams per hectare in some cases). Their future development will be followed with interest.

Oils

Mineral oils in emulsion form were formerly used widely for winter spraying of deciduous fruit trees to kill the overwintering stages of pests. This particular practice has largely disappeared with the introduction of improved insecticides for summer season use, but oil sprays are still valuable for control of certain pests such as scale insects against which many insecticides are ineffective. Oil sprays kill insects by a physical smothering action which interferes with respiration. Many different oil based preparations have been marketed and it is important that the most suitable be selected for a particular use in order to obtain maximum effect and at the same time avoid plant injury. Some plants are sensitive to oil sprays and should not be treated with them. Detailed advice should be sought.

Insect growth regulators

Considerable interest has been aroused in recent years in the possibility of developing insecticides which depend for their activity on interfering with insect growth hormone systems. It has been known for some time that insects can be killed experimentally by treating them with substances which mimic the activity of insect juvenile hormone or of moulting hormone, provided that treatments are applied at a critical stage of development. High dosages were required with earlier materials but much more potent compounds have now been discovered. Some are effective enough to market as insecticides and the first of these is now registered for limited purposes in North America. None has so far been approved for use in Great Britain. Some of these compounds may kill insects other than by straight juvenile or moulting hormone activity and a new group name of **insect growth regulators** (IGRs) has been coined which includes chemicals with such properties as well as insect hormones. It may be expected that further commercial products based on such substances will appear in the future.

Miscellaneous compounds

Besides the main groups of insecticides described above many other synthetic chemicals have been evaluated or used in practice for pest control. This is particularly the case with materials for control of mites (miticides), many of which do not fall into any of the above named chemical categories.

Insecticides and miticides, for which there are currently approved products in Great Britain are listed in Appendix 1 together with their main trade names, mammalian toxicity, and main uses.

(d) Nomenclature

The full chemical names of insecticides are complex, unwieldy and unlikely to be meaningful to the average person. For instance, the full chemical name for the insecticide commonly known as diazinon is diethyl 2-isopropyl-6-methyl-4-pyrimidinyl phosphorothionate. Insecticides are therefore given common names (diazinon, DDT, carbaryl, for example) which are for the most part internationally standardised. Some commercial products include the common name in the trade name but more often separate trade names are coined. If there is more than one manufacturer there may be several trade names for products with the same active ingredient. This can be confusing to the user, but in all cases it is a legal requirement for the common name to appear on the label together with the percentage concentration of the active ingredient. (For further details of labelling requirements see later under Registration and Labelling, page 214.)

One weakness of the use of common names for insecticides is that many compounds occur in a number of isomeric forms which may differ widely in insecticidal activity. The most active isomer of DDT for instance is pp′DDT and the quality of a commercial DDT preparation depends on the percentage of this isomer present. In some instances therefore the common name is inadequate without further chemical detail.

Excellent information on nomenclature of agricultural chemicals (and on many other technical aspects) is provided in the "Pesticide Manual" published by the British Crop Protection Council and regularly updated.

(e) Mode of action

In considering the mode of action of insecticides two separate aspects can be identified. One concerns the way in which the insect acquires a dose of the chemical, the other the actual mechanism by which the insecticide interferes with some vital bodily process.

Before an insecticide can exert any effect it must penetrate the intact insect from the outside. Four routes of entry are recognised and insecticides may be grouped according to the main route by which they are taken in.

Contact insecticides are those that have the ability to penetrate the intact body surface after it becomes contaminated with a spray droplet, or by the insect walking over a surface on which there is a deposit of the insecticide.

Stomach poisons on the other hand must be ingested and are then absorbed from the gut into the general body system. In practice stomach poisons are usually applied to the insect's normal food (such as leaves of a host plant for a plant feeding caterpillar) but may be incorporated into special bait preparations.

Fumigants are insecticides which are sufficiently volatile to produce a toxic concentration in gaseous form in the air around an insect. Entry is primarily through the insect's spiracles. True fumigants are gases at normal temperatures and pressures and must be stored in pressurised cylinders. Methyl bromide is an example. Such fumigants are rarely used to control pests on growing plants, but are often used to disinfest stored grain and other similar materials. However, many other insecticides which are normally applied as sprays are sufficiently volatile to exert considerable fumigant action and thus are able to kill insects at a distance. HCH and dichlorvos are examples. Fumigant activity is a valuable property when controlling "hard to get at" pests (such as aphids within curled leaves), as direct contact of the insect with the spray droplets is not essential.

Systemic insecticides have the property of being absorbed into the sap stream of plants (from leaves or roots) and then of being translocated to other parts of the plant. Movement is almost entirely upward, as this is the main direction of flow of plant sap. As uptake of the insecticide by insects is dependent on the ingestion of plant sap, most systemic insecticides are effective only against sap feeders (Hemiptera primarily). However, some newer insecticides which are systemic in action also have some effect on biting/chewing insects. A few insecticides of low mammalian toxicity can act in a systemic fashion in higher animals by being absorbed into and carried by the blood stream.

The above terms usefully describe modes of uptake of insecticides but they do not provide rigid categories. A particular insecticide for instance may show contact activity but also be systemic within the plant, while another insecticide may be a stomach poison but also act on contact. In the selection of an insecticide for a particular purpose knowledge of these properties is extremely important if best results are to be obtained.

Most insecticides poison insects by interfering with the functioning of the nervous system, resulting initially in increased uncoordinated muscular activity but leading eventually to paralysis and death. With organophosphate and carbamate insecticides this is brought about because these substances inhibit the enzyme cholinesterase. The function of this enzyme is to break down acetylcholine which is formed at nerve connections (synapses) each time a nerve impulse is transmitted. If cholinesterase is inactivated, acetylcholine accumulates and the nerve ceases to function properly. Recovery can occur from mild poisoning but at higher dosages death ensues. Organochlorine insecticides also affect the nervous system but they are not cholinesterase inhibitors and exactly how they work is not clear.

Some newer insecticides have completely different modes of action. Diflubenzuron (trade name Dimilin) for instance disrupts moulting and the insect dies without successfully completing the process. Between moults it is unaffected. Other new insecticides (already referred to as insect growth regulators) interfere with the insect's growth hormone systems resulting in abnormal development and death. Further introduction of insecticides with other novel modes of action can be expected.

(f) Spectrum of activity

The range of pests against which an insecticide is active is known as its **spectrum of activity**. Some insecticides such as DDT are effective against pests from many different groups and are therefore **broad spectrum**. In other cases (**narrow spectrum insecticides**) only a limited range of insects is affected. The carbamate insecticide pirimicarb falls into the latter category as its activity is primarily against aphids. Although broad spectrum insecticides are useful from the point of view of being effective against several pests at one time, selectivity is desirable to minimise harmful effects on beneficial species. No completely selective insecticides are known but some have far less effect on beneficial insects than others and their use is particularly important in pest management programmes (see Chapter 12). It probably is technically possible to produce insecticides more specific in action than many at present in use but this is unattractive to insecticide manufacturers as costs of discovery and development are similar to broad spectrum materials but potential sales are much less.

The term "**non-target organism**" is sometimes used to refer to species other than the pest against which insecticide treatment is aimed. The possible kinds of organisms included can be extremely diverse. The most important are (a) parasites and predators of the pest to be controlled (or of potential pests), (b) pollinating insects, (c) components of the soil fauna such as earthworms, and (d) wildlife (birds, mammals and fish). Information on how hazardous a new insecticide is to these groups is required before approval is granted.

Many insecticides are highly toxic to bees and very few can be safely applied to crops in flower. This presents problems in chemical control of pests which attack flower parts or developing seed and careful selection of insecticides is necessary if they are to be used for such purposes. Some short lived materials are reasonably safe if they are applied late in the day after bees have finished foraging, since any toxic effect will have largely disappeared by morning. Detailed advice should always be sought if insecticides are to be applied to crops in flower.

(g) Persistence

Immediately an insecticide is applied its activity starts to decline. Insecticides whose activity declines slowly are said to have **long persistence** (or simply to be persistent) while those whose activity disappears quickly have **short persistence** (are transient). The degree of persistence varies greatly between different insecticides and to some extent for the same insecticide under different conditions. On plant surfaces effective persistence varies from a matter of hours to several weeks.

Degree of persistence is determined by several factors. Firstly there is the chemical stability of the insecticide itself. Some, such as DDT, are extremely stable and break down only very slowly in the environment, but others decompose rapidly. Chemical breakdown is usually accelerated by moisture, sunlight and high temperatures. Secondly, under open field conditions, insecticides may be washed from plant surfaces by rainfall. Chemicals with high water solubility will obviously be most affected. Thirdly, as plant surfaces grow, any deposit of insecticide will become more spread out and diluted. Fourthly, many chemicals are volatile to some extent so that they slowly (or in some cases quite rapidly) evaporate into the atmosphere. Finally, there is the possibility of chemical breakdown brought about by micro-organisms. This is a major factor affecting the life of pesticides in the soil.

(h) Toxicity (to higher animals)

Virtually all insecticides are to some extent toxic to higher animals, including man, as well as to the pests against which they are applied, but the degree of mammalian toxicity varies greatly from one insecticide to another.

Registration of a new insecticide requires the submission of extensive information concerning toxicity to higher animals. This obviously cannot be obtained by experiments on human beings so that tests are carried out on laboratory animals the results of which enable assessment to be made of likely toxicity to humans. There is danger in such extrapolation however because human beings may be more (or less) sensitive than the animals on which tests were conducted. A large safety margin is therefore employed in assessing likely hazards to humans.

The main laboratory animals used for toxicity studies are rats but tests are also often carried out on rabbits (especially for assessing dermal toxicity) and on dogs (especially for long term feeding studies). No single toxicity test can provide all of the information required so several kinds of tests are always undertaken:

Acute oral toxicity — To evaluate acute oral toxicity the material is fed to the test animals in their normal food, and mortality and sub-lethal effects measured over a short period of time. Results are usually expressed as the dosage (in mg per kg body weight) of chemical required to kill 50 percent of the test animals (LD_{50}). The *lower* the LD_{50} value the *more toxic* is the substance because a smaller amount is required to produce the same effect compared to a larger amount of a less toxic substance. An insecticide with an acute oral LD_{50} of 10 mg/kg is thus very toxic whilst an insecticide with an LD_{50} of 1,000 mg/kg is only slightly toxic.

Dermal toxicity — In this case the insecticide is applied to the intact (but shaved) external body surface in an organic solvent or oil. Effects are assessed after a fairly short period of time, as with acute oral toxicity evaluation. Dermal toxicity tests may also involve application directly into the eye as this is a very sensitive part of the body and easily contaminated by splash in practical use of pesticides.

Inhalation toxicity — When applying insecticides, fine spray droplets, dust particles or the insecticide itself in vapour form may be inhaled. Toxicity testing may therefore also include this method of exposure.

Chronic toxicity — All of the above types of tests involve evaluation of effects a comparatively short period of time after administering a single dose of insecticide, or after brief exposure. There is also the possibility however that a chemical will show toxic effects only following continual exposure to low levels over long periods of time. Toxicity testing therefore also involves continual feeding studies of low dosages to laboratory animals for at least two generations. The test animals are examined for sub-lethal effects including blood cell counts and changes in liver tissue. Also included in such chronic toxicity studies are evaluation for possible **carcinogenic** (cancer inducing) and **teratogenic** (producing birth deformities) properties.

From an accumulation of data from all of the foregoing tests assessment is made of likely hazards to humans from a new pesticide. Only when authorities are satisfied that such hazards are within acceptable limits is permission given for the material to be marketed.

The hazard involved in the use of an insecticidal formulation is not the same thing as the toxicity of its active ingredient. Toxicity of an insecticide only varies to any extent with the method of exposure to the substance (oral, dermal, inhalation) but degree of hazard depends on a number of factors, in particular the type of formulation, percentage active ingredient, application equipment employed and circumstances in which the material is used (eg, under glass or out of doors). Granular formulations for example are safer to handle than wettable powders or emulsifiable concentrates because they are of lower percentage active ingredient and do not have to be measured or diluted before application. Similarly there is less risk of inhalation when applying insecticides out of doors compared to under glass.

In granting official approval to an insecticide product the authorities may prescribe (a) conditions of sale and of use (to protect those handling and applying the material), (b) on what plants and time of application in relation to harvest (to protect consumers) (see under Registration and Labelling p. 214) and (c) precautions to be taken in handling and applying the insecticide including use of appropriate protective clothing. All of this and other pertinent information is included on the label.

(i) Ecological hazards

In addition to the above toxicological information, the intending registrant of any new insecticide must submit results of detailed studies to show what happens to the chemical in the environment. This information must cover such aspects as pathways and rate of chemical breakdown in plants and soil, movement within the environment, and possible effects on non-target organisms including wildlife. When a material is finally approved for sale a great deal of information is therefore available as to what its overall environmental effects are likely to be as well as effectiveness for the purpose(s) claimed.

(j) Phytotoxicity

Phytotoxicity means toxicity (injury) to plant life. Obviously a potential insecticide that was highly injurious to plants would never be developed for crop use, but sometimes an insecticide that is generally safe may be damaging to just a few species of plants, or to certain cultivars. Additionally some formulations (see below) may be more hazardous than others. It is important therefore, to be aware of any phytotoxic hazard when using an insecticide. Such information is included on the label.

(k) Formulation

Very few insecticides are suitable for use in the raw chemical state. They nearly always require to be diluted with a suitable inert material and to have other substances added to impart particular properties to the final product; that is, the insecticide has to be formulated. There are several different kinds of pesticide formulations designed for different uses. The main types are:

(1) *Emulsifiable concentrate (EC)*

The insecticide is dissolved in an organic solvent to which emulsifiers are added. When the concentrate is added to water a milky emulsion is formed.

(2) *Wettable powder (WP)*

A wettable powder consists of very finely ground particles of a suitable inert carrier (such as a clay) to which the insecticide is added. Small quantities of surfactants* are also incorporated so that the powder readily wets and disperses in water. In diluted form wettable powders consist of small solid particles in suspension. These tend to settle out if left to stand so that wettable powders should be agitated continually in the spray tank or applied immediately after dilution.

(3) *Soluble powder*

A few insecticides (eg trichlorfon) are sufficiently water soluble to enable them to be applied as true solutions. Suitable surfactants are usually included in the formulation to improve spreading and adhesion to plant surfaces.

(4) *Dust*

Dusts are prepared in a similar manner to wettable powders by adding the insecticide to, and grinding it with, a suitable inert diluent. However, in contrast, dusts are applied dry without dilution. The percentage active ingredient in dust preparations is usually low (1-5%). Dusts do not adhere well to plant surfaces but may be useful in dry climates where water for spraying is scarce or in situations where access for spraying is difficult, eg for blowing into wall cavities for cockroach control.

(5) *Granule*

Granule formulations consist of relatively large (about 1-3 mm diameter) particles of a carrier with a low (5-10%) concentration of insecticide. The granules are formed from either a clay type material into which the insecticide is absorbed and which disintegrates to release it, or from a harder substance such as calcite or

* Substances with surface active properties, such as wetting agents.

silica on which the insecticide is coated. In the latter case the coating strips off to release the insecticide. Granular formulations are extremely useful for certain purposes such as application of systemic insecticides to the soil at sowing time. Slow release of the insecticide from the granule provides protection of the young plant for the first few weeks of its life. Because of their large particle size there is little or no drift when using granules.

(6) *Bait*

The idea of incorporating an insecticide into a food based bait which a pest will consume is a good one. It is usually extremely economical, as most of the insecticide reaches its target (the pest) and little is wasted, and also effects on non-target organisms are likely to be minimal. An insecticide for use in a bait must have good stomach poison action (as most will be ingested) and must not be repellent to the pest to be controlled. In practice there are a number of problems in developing effective bait formulations for control of crop pests. One problem is to find suitable food bases for the bait which the pest prefers compared to living plant material that usually is simultaneously available. Another problem is that some bait mixtures have to be freshly prepared for best effect. They may also be difficult to apply evenly in the field. If effective chemical attractants (see later in this Chapter) can be found for pests, bait preparations could be considerably improved.

(7) *Specialised formulations*

The above does not exhaust completely the different types of insecticide formulations that are available. There are many others for specialised purposes. One of the most familiar is the aerosol spray canister for household use. This consists of an insecticide mixture in a volatile solvent which is expelled through a fine nozzle by a gas under pressure (propellant) to produce a fine spray.

For glasshouse pest control, or for use against furniture beetle, smoke generators are highly convenient. The insecticide in this case is incorporated into a pyrotechnic mixture which produces a dense smoke (including the insecticide) when ignited. Penetration throughout the area treated is good but as near airtight enclosure as possible is essential for best results. Only insecticides that can withstand the heat of combustion without undue chemical breakdown can be formulated in this way.

(1) Compatibility

Where more than one pest has to be controlled at a time, or a pest and a plant disease have to be dealt with simultaneously, it is common practice to mix two (or more) pesticides in the spray tank and apply them together. If two materials can be safely mixed in this way without untoward effects (lack of control, plant damage) they are said to be **compatible**. However, all possible combinations are not safe as there may be reaction between active ingredients or between formulations. Such mixtures are **incompatible**. Chemical manufacturers often provide **compatibility charts** which indicate those materials that can be safely mixed together and those that cannot. Anyone planning to apply a combination of pesticides should check to ensure that the proposed mixture is compatible.

(m) Application equipment

Equipment used to apply insecticides must be suited to the crop (or other situation in which the pest is to be controlled), to the scale of the operation, and to the formulation of chemical to be applied. There are three broad groups of application equipment (sprayers, dusters and granule applicators) with many variations of each. In addition, specialised equipment of various sorts (such as thermal fog generators) is available for particular purposes.

Detailed consideration of pesticide application equipment would require a book to itself so that here brief consideration only is given to some important features of the main types of equipment and to factors which are important in their operation. For details of application equipment best suited to particular purposes and of its operation the reader should consult suppliers of agricultural chemicals and spray machinery.

(1) *Sprayers*

Sprayers vary greatly in size from small hand operated garden syringes to huge orchard air blast machines (mist blowers) that require a large tractor to tow them. However, all sprayers must provide some means of breaking up the spray liquid into droplets (**atomising**) and of conveying the droplets onto the object to be sprayed (**target**). In most smaller machines hydraulic pressure provides the means both of atomising the spray liquid and of imparting momentum to the droplets so that they carry to and impinge on the target. Many larger machines however, especially those designed for low volume application (see below), employ a fan generated stream of air to carry the droplets. They are referred to as **air assisted** or **air blast sprayers**. Atomisation in such sprayers may be through normal hydraulic nozzles or by power operated spinning discs or spinning cages, which provide more uniform droplet size than conventional nozzles. Until about 25 years ago all sprayers were of the hydraulic type and applied pesticides in very dilute form. The objective of spraying with them was to completely wet plants (or other target surfaces) normally to the point of run off. This is still the procedure with small hydraulic sprayers but the development of air blast machines has permitted the use of chemicals in more concentrated form. The spray pattern is then one of small discrete droplets which do not give complete cover of the target or immediately coalesce but nevertheless provide effective control of most pests. This is **low volume spraying** compared to earlier high volume application. Most large modern sprayers are of the low volume type.

Further refinement of equipment has enabled spray volumes to be progressively reduced and the extreme is reached in **ultra low volume** (ULV) spraying where the insecticide is applied in extremely concentrated form (50% or more) or even as the undiluted technical material if this is a liquid (eg, malathion). With ultra low volume application *very* small droplets and precise application are essential for effective results.

There is no general agreement as to where dividing lines should be drawn between high volume, low volume and ultra low volume but as a general rule high volume spraying involves the application of 1120 litres or more per hectare (100 gals + per acre), low volume 225-560 litres per hectare (20-50 gals per acre) and ULV 50 litres per hectare (5 gal per acre) or less.

Emulsifiable concentrate or wettable powder formulations (diluted with water) are used for high volume or low volume spraying but ULV usually requires special formulations with a light oil as diluent.

Low volume (and ultra low volume) spray techniques require less frequent filling of the spray tank compared to high volume spraying. This is very important in aerial spraying and can materially reduce application costs. Also, compared to high volume spraying, less pesticide is normally used to achieve the same result as more is deposited on target. On the other hand more precision is required with low volume spraying in setting up and operating the equipment. This applies even more to ULV. Even with the best available equipment much spray is wasted when treating growing plants (as much as 70 per cent commonly does not reach the target) so there is still great scope for improvement in the efficiency of spray equipment.

Coverage

An important aspect of spraying is adequate coverage. Three types, or levels, of cover can be distinguished. **Ground cover** (Fig 27(a)) refers to how well the spray equipment covers the ground

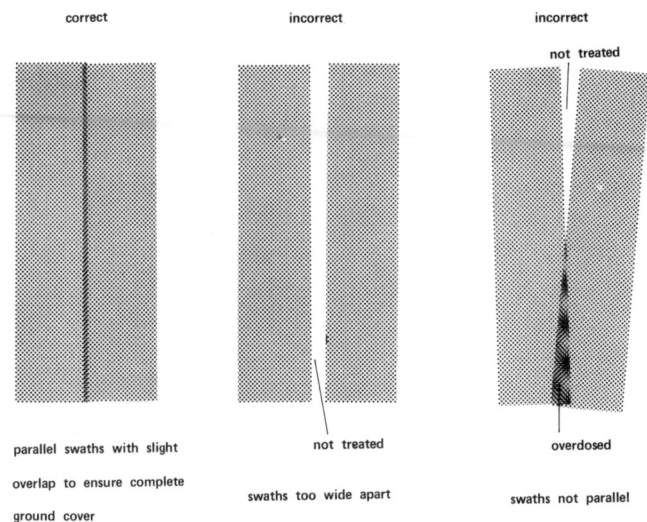

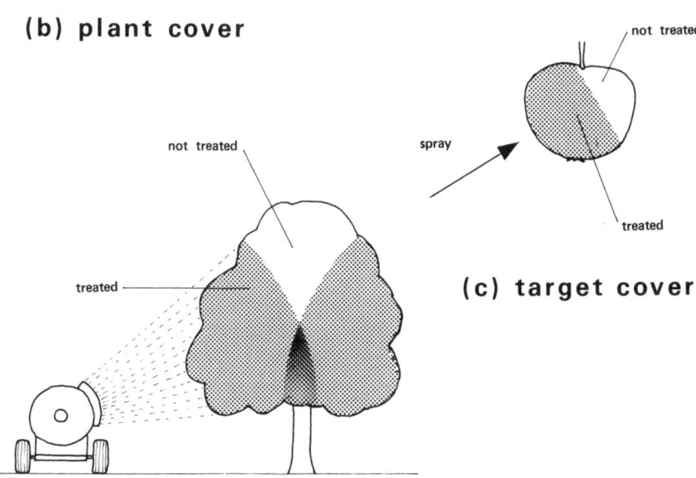

Fig. 27. Spray coverage.

area being treated. Ideally swaths (paths treated by one pass of the sprayer) should be parallel and slightly overlapping. Deviations from this (see Fig 27) will result in incomplete coverage and/or overdosing of some areas. **Plant cover** (Fig 27(b)) refers to the degree of coverage of individual plants within the area sprayed. Figure 27(b) indicates that complete cover of large targets (eg mature fruit trees) is difficult to achieve. **Target cover** (Fig 27(c)) refers to coverage of the crucial portion of the plant with respect to a particular pest (for example apple fruitlets in the case of codling moth). Again complete cover is often difficult to attain.

Rates of use

Where high volume spraying is carried out the amount of insecticide required may be stated as so much percentage active ingredient (a.i.) in the dilute spray. A common value for most insecticides would be 0.1 to 0.05%. However such a mode of expressing rate of use is not suitable for low volume (including ULV) application as greatly different amounts of spray liquid may be applied per hectare depending on equipment. It is therefore necessary to state the amount of active ingredient required per unit area of ground (per hectare, or per 10 square metres for example). Rates of use in these terms for many insecticides would be 1 or 2 kg a.i. per hectare (kg/ha). It is largely immaterial how much water is used to dilute and apply this amount of chemical provided adequate target cover is achieved.

(2) *Dusters*

Dusts are used far less than sprays as a means of applying insecticides to plants and so there has been less development of dusting equipment. Most dusters incorporate a metering device of some sort to regulate dust output and a blower to carry the dust to the target. Attempts have been made to charge dust particles electrostatically so that they adhere better to plant surfaces but this has not been very successful in practice.

(3) *Granule applicators*

Various types of granule application equipment are available. Selection for a particular purpose depends on whether the granules are to be broadcast (eg, over pasture) or applied in a restricted manner (eg, in furrow application at sowing time). Most granule applicators incorporate a mechanism to feed granules in regulated amounts from the hopper. For in furrow application, granules may be simply released "down the spout" into, or adjacent to, the seed furrow. For broadcast application there is usually some means of throwing the granules to form a swath. Common mechanisms are a rapidly rotating ridged disc or oscillating arm.

(n) Resistance

Insecticide resistance may be defined as the developed ability of an insect population to withstand an insecticide which was formerly effective. Two points should be noted from this definition. Firstly, resistance is developed. It did not previously occur, so this implies some change in the insect concerned. Secondly, the term insect population is used rather than insect species, because resistance normally arises only in certain areas of a species' distribution. It may however eventually become widespread.

Although insect resistance to insecticides has only become important during the past 20 years or so, it is not by any means a new phenomenon. The first recorded case was at about the turn of the century and concerned a scale insect resistant to hydrogen cyanide. What is new however is the widespread nature and seriousness of the problem. Initially regarded as something of a curiosity, resistance to one or more insecticides now occurs in several hundred agricultural and public health pests throughout the world. Its occurrence in all cases is related to the intensity of insecticide use, for reasons that are explained below. There are exactly parallel situations of drug resistance among micro-organisms causing disease in humans (such as the malarial parasite) and to anthelminthic drenches among animal parasites. Resistance of plant disease organisms to chemicals used for their control has been much slower to develop and was unknown until a few years ago. Unfortunately cases are now being reported more frequently, particularly to systemic fungicides.

Resistance to insecticides unfortunately seems to be the rule wherever suppression of a pest species is attempted with the same chemical for a sufficient period of time. It is a striking example of evolutionary adaptation which enables living organisms to survive in the face of threats to their existence.

Detection of resistance

The failure of a chemical treatment in the field to control a pest is *not* adequate proof of the existence of resistance, even though satisfactory control may have been previously achieved with the insecticide in question. It is however a pointer that resistance may be involved and should be investigated. Many other reasons can account for control failures with insecticides, such as change to a new formulation or change in application equipment (from high volume to low volume spraying for instance) resulting in inadequate coverage. There is also the possibility that abnormal environmental conditions may be involved. If such alternatives can be ruled out resistance is more likely to be the cause. Final proof can only come from laboratory tests on samples of the suspected resistant insects. Such testing involves comparison with a known susceptible strain of the same species. The procedure must be standardised for consistent results. Changes in susceptibility of juvenile insects with growth is an added complication that must be taken into account. Recommended test procedures for detecting resistance have now been published for many important pests. Results of such tests are usually expressed as the ratio of LD_{50} values (dosage to kill 50 percent of test insects) between suspected resistant and known susceptible samples. This ratio should be at least x10 before resistance can be confidently stated to be present. In practice, values of x100 or even x1000 are not uncommon.

Resistance mechanisms

To ask what is the mechanism of insect resistance really involves two separate questions: firstly, how does resistance arise? secondly, how do resistant insects differ from susceptible ones?

The answer to the first question is clear cut. Resistance arises through a selection process operating over a number of generations. Selection results from exposure to the insecticide so that susceptible individuals are killed while the more resistant ones survive and reproduce. After this has been repeated over several generations the population consists mostly of resistant individuals. The fact that resistance is passed on from parents to offspring indicates that it is genetically determined. There is no change in susceptibility of an *individual insect* during its lifetime. Resistant insects are no different in appearance from susceptible ones. The only difference is their ability to withstand the insecticide.

The second question (how do resistant insects differ from susceptible ones?) refers not just to the fact that resistant insects can withstand the insecticide in question, but rather to exactly how they do this and avoid being poisoned compared to susceptible ones. This aspect has received a great deal of research attention, and we probably now know more about how resistant insects withstand insecticides than we do about how susceptible ones are killed!

Several resistance mechanisms have been identified.

(1) *Detoxication.* This involves the ability (of the insect) to chemically modify (detoxify) the insecticide quite rapidly so that it loses its toxic action. This is the most important and widespread resistance mechanism. In several cases specific enzymes have been

indentified which bring about the necessary chemical reactions.

(2) *Insensitive target.* The principle here is different in that the site of action (target) of the insecticide is no longer affected, or at least to a much lesser extent. The insecticide is broken down no more rapidly than in susceptible insects. Some cases of resistance to organophosphorus insecticides are known to operate in this way.

(3) *Slower rate of penetration.* All insecticides must penetrate the insect body (either from the outer body surface or through the gut wall following ingestion) before they exert any toxic action. If the rate of penetration is reduced sufficiently the insect may be able to detoxify the insecticide rapidly enough to prevent itself being poisoned.

(4) *Storage.* In early investigations of insecticide resistance it was suggested that resistant insects might store insecticide in certain parts of the body (such as fat body) so that toxic action was reduced. This may be a contributing factor in some cases but is not an important one.

(5) *Avoidance.* If an insect can avoid insecticide treated surfaces it will not be killed even though it is unable to withstand the insecticide should it come into contact. Such **behavioural resistance** is known to occur in some insects (eg, some mosquitoes) and is quite different from the other mechanisms described. It cannot be detected by normal testing procedures and requires special techniques for its evaluation.

There are therefore several resistance mechanisms known all of which have been proven with one insect species or another. There is nothing to prevent more than one mechanism operating simultaneously in the same insect and this is known to occur in some cases.

Possible countermeasures

One possible means that is sometimes suggested for dealing with resistance is to increase the rate of use of the insecticide. This is impractical (except at very low levels of resistance) because it would require many times the normal quantity of insecticide to be effective. Such rates of use are uneconomic apart from other considerations such as environmental contamination.

Another idea that has been proposed and researched is to try to find chemicals which, when added to the insecticide, restores its effectiveness. Such substances are known as **synergists** and a number have been discovered. They seem to work by inhibiting enzymes which detoxify the insecticide (see previous section). As this is not the only possible mechanism of resistance their application is limited, even if good synergists are available. A further problem is that after a while insects seem able to become resistant to the synergist/insecticide mixture so that there may be no more than a temporary respite.

The only practical answer so far to resistance is to change to another insecticide with a *different mode of action*. Unfortunately, although there are many individual insecticides, they fall into a very limited number of chemical groups, and within each group modes of action are the same or very similar. The choice of alternative insecticides is therefore quite limited. In any case it is usually only a matter of time before resistance develops to the second chemical so that a third must then be found. This process cannot continue for long before all possible alternatives have been used and this is now the situation with some pests.

It might be thought that if three or four insecticides were used in succession, by the time the usefulness of the last one had been exhausted susceptibility to the first might have returned. To some extent this does occur but once the first used insecticide is employed again resistance to it develops much more rapidly than it did initially, so that this practice is of little value.

Conclusion

Resistance to insecticides is a very serious problem threatening the continued effective control of many important pests. No satisfactory answer to the problem has yet been found, and may not even be possible in purely chemical terms. Permanent solutions must be sought by adopting a broader ecological approach to pest problems involving less intensive use of insecticides (see Chapter 12).

(o) Residues and tolerances

The amount of insecticide on a plant surface or other substrate immediately following application is known as the **initial deposit** (or simply as the deposit). It declines with time due to chemical breakdown, volatilisation and other weathering processes. That which remains on the harvested part of the plant at crop maturity is referred to as a **residue** and is usually expressed in parts per million (ppm) fresh weight of produce. Residue levels at harvest vary enormously depending on such factors as stability of the insecticide, the type of formulation used, climatic conditions and the crop concerned. Some insecticides decline very rapidly to below detectable levels but others may persist for long periods. Residues of some insecticides remain entirely on the outside of the plant but residues of others (particularly systemic insecticides) may occur partly inside fruits and other harvested parts.

Because insecticides are toxic to humans, residues in the harvested parts of food crops must be carefully controlled. To this end most countries have established maximum permissible levels for different insecticides in harvested produce. These are known as **tolerances**. The level at which they are set depends on the toxicity of the insecticide and on the importance of the food item in the national diet (because this will determine likely daily intake). Tolerance levels thus vary somewhat between different commodities and between countries. However, in all cases there is a large inbuilt safety factor to protect the consumer.

It is one thing to define tolerance for insecticides in food crops but how does the grower or farmer know that insecticide residues in his produce will not exceed permitted levels? This is achieved by prescribing minimum periods (**waiting periods**) between last application and harvest. These vary according to the insecticide (and the crop) and are part of essential information included on the label. Similar periods are laid down following insecticide treatment of pastures and fodder crops during which stock must be kept out of the treated area.

(p) Attractants, repellents and pheromones

An **attractant** is a chemical substance towards which insects make orientated movements. Attractants influence insects from a distance and must therefore be volatile to some extent. A **repellent** in contrast exerts the opposite effect in that it induces insects to move away from its source. It too must be volatile. **Pheromone** is a fairly recently coined word to describe chemicals produced by insects (and some other organisms) to convey a message to other members of the same species. Such substances must also be volatile to some degree. Many pheromones of insects are in fact attractants (particularly the sex attractants — see Chapter 5 and later in this section) but others convey an entirely different message. The trail marking pheromones of ants for example induce worker ants to follow a trail, while the alarm pheromone of aphids causes the insects to drop from the plant. Such powerful behaviour modifying substances obviously offer considerable potential for manipulation of pest species, a potential which is only just beginning to be exploited.

(1) *Use of insect attractants (other than sex attractants)*

Chemical attractants for particular species of insects have been discovered either by analysis of materials (naturally occurring or man made) which have been observed to be attractive, or by chemical synthesis and testing. Insects are likely to respond to attractants for one of two reasons; either for the purpose of feeding (in which case both sexes will probably be attracted), or for egg laying, when only gravid females will respond.

One obvious use of chemical food attractants is for addition to poison baits to improve their effectiveness. Many insecticide bait formulae therefore include an attractive substance of some sort. Chemical insect attractants may also be used in conjunction with mechanical trapping devices but such use is expensive and usually not particularly effective for control purposes.

(2) *Use of insect sex attractants*

Insect sex attractants are pheromones produced by one sex to attract the opposite sex for the purpose of mating. They are active at very low concentrations and are highly specific for the species concerned. Although they are produced by females and attract only males (with few exceptions) their great potency and selectivity offer considerable potential for the manipulation of pests. Many insect sex attractants have now been synthesised and some are available in commercial form. There are three main ways in which they may be used to assist with control of pests.

(i) *For incorporation in traps or poison baits*. If traps or baits incorporating an insect sex attractant are to provide control of a pest species two conditions must be satisfied. Firstly, the trap or bait must compete successfully with natural pheromone emitting females of the species. Secondly, a very high proportion of males must be eliminated from the population before they are able to mate with females if there is to be effective suppression of the next generation. There are few examples so far where success has been achieved with sex attractants used in this manner.

(ii) *To disrupt mating*. The idea here is that, if a sex attractant is released into the atmosphere and maintained at sufficient concentration over the period of mating of a species, the natural scent of females will be swamped and males will thus be unable to locate them. The procedure is sometimes referred to as the "**male confusion technique**". Success has been achieved experimentally with the technique for several species of moths but how practical it will prove to be for general use is as yet uncertain. There are many difficulties involved such as providing for adequate distribution of the pheromone and maintaining high enough concentrations for sufficient periods of time. Cost is also likely to be a major factor. Despite such problems research into this use of sex attractants is likely to continue because it offers the possibility of very selective pest control with minimum disturbance of other organisms.

(iii) *For survey and monitoring*. The specific and potent properties of sex attractants make them ideal for use in insect traps to survey for the presence of particular pests on an area basis and to monitor for seasonal occurence. They are already being extensively used for these purposes and commercial supplies of the sex attractant, together with suitable traps, are available for several important pests such as codling moth.

(3) *Repellents*

Chemical repellents have been employed for many years for personal protection against blood sucking insects. A well known example is DIMP (dimethylphthalate) used for protection against black flies and mosquitoes. However, repellents have never been successfully developed for plant pest control, one of the problems being that continual emission is necessary for effective protection.

(q) Feeding deterrents

A further class of chemicals that influence insect behaviour are **insect feeding deterrents**. They differ from repellents in that feeding is suppressed only after contact is made with the chemical. Strictly speaking there is no action at a distance so that deterrents are not volatile but it is of course possible for a chemical to act in both ways to some extent. The principle of controlling plant pests by suppressing feeding has considerable appeal because natural enemies would be completely unharmed. Plant constituents with feeding deterrent properties are a major factor in the resistance of plants to pests.

Considerable research has been conducted into the possibility of protecting plants from pests by treating them with synthetic feeding deterrent chemicals. Although some successes have been reported the approach has never been successfully developed for practical use. One problem is that new growth of the plant is unprotected. Feeding deterrents with systemic properties would help to overcome this if they could be developed.

(r) Chemosterilants

There are several ways of inducing sterility in insects, one of which involves the use of chemicals (**chemosterilants**). These are discussed briefly here as part of chemical control procedures. The advantages associated with sterilising a pest insect rather than killing it outright are discussed later in this Chapter where the sterility principle in pest control is considered.

Many chemicals are known which have the ability to sterilise insects. Some must be ingested but others are active by external contact, including in some cases uptake by insects walking over a treated surface. Treatment early in adult life before the insect is sexually mature is usually necessary. Some chemosterilants can be applied to the larval or pupal stages which then produce sterile adults. Unfortunately all insect sterilants to date are hazardous to humans, some extremely so because of their carcinogenic properties, and this is a major limitation which has so far restricted their practical development.

(s) Registration and labelling

In most countries any chemical substance used for the protection of plants (or animals) against pests or diseases, or for control of weeds, must be officially registered before it may be offered for sale. In the United Kingdom the use of pesticides is similarly regulated by two schemes agreed between government and industry; the Pesticides Safety Precautions Scheme and the Agricultural Chemicals Approval Scheme.

*The Pesticides Safety Precautions Scheme**

The primary objective of this scheme is to ensure that users of pesticides are adequately protected against any toxic hazards that may be involved in handling and application. There is also responsibility to safeguard livestock, domestic animals, beneficial insects, wildlife and the environment in general. The scheme covers all pesticides (insecticides, acaricides, fungicides, herbicides, rodenticides and plant growth regulators) whether used in agriculture, horticulture, forestry, food storage or public hygiene.

Under the scheme chemical manufacturers give official notification before they intend to market a new product, extend uses of an existing product, or change a formulation or method and rate of application. Four categories of clearance are provided ranging from Trials Clearance for one season only without sale of the product, to Commercial Clearance which allows the product to be sold on the open market subject to agreed labelling and any requirements concerning conditions of use such as specified protective clothing, rate of use, crops treated and time of application in relation to harvest.

The Agricultural Chemicals Approved Scheme

This scheme provides a means of granting official approval for the

* — The Pesticides Safety Precautions Scheme is supported by legislation such as the Poisons Act 1972, which governs labelling, storage and sale of scheduled poisons and the Health and Safety (Agriculture) (Poisonous Substances) Regulations 1975 which specify conditions of use and protective clothing required for applying scheduled substances.

efficacy of products that have already been cleared under the Pesticides Safety Precautions Scheme. The present scheme dates from 1960 but grew out of an earlier Crop Protection Products Approval Scheme which originated in 1942.

Approval is based on data submitted by the manufacturer and on inspection of field trials by officers of the scheme. Valid trials from at least two seasons are normally required together with information on composition of the product, application properties, storage life, biological efficacy and crop safety. Products which have been approved have a distinctive emblem incorporated on the label. A list of "Approved Products for Farmers and Growers" is published annually.

Although products are approved under the scheme rather than active ingredients, a list of insecticides/acaricides for which there are currently approved products is provided in Appendix I for reference.

Part of the approval process involves approval of the label which must go on all containers in which the material is sold. Amongst other information, the label must state the name and percentage content of the active ingredient, the poison classification, and precautions to be observed in handling and applying the material. The label must also precisely state directions for use and pests controlled. The label of an insecticide product is therefore an extremely important and informative document. The single most important instruction that can be given to the user of pesticides is to *read and follow the label.*

PLANT AND ANIMAL QUARANTINE

Although human activities have greatly assisted the spread of pest species around the world this process is fortunately by no means complete and for most countries there is a long list of plant and animal pests (and diseases) which have not yet gained entry. This is particularly the case with geographically isolated countries such as New Zealand. The aim of plant and animal quarantine is to prevent the introduction and further spread of pest and disease organisms. It cannot be undertaken by individuals but must be the responsibility of governments to establish suitable legislation and to provide for its implementation. The brief discussion that follows concerns the principles and objectives of quarantine which are the same everywhere. Details of legislation however vary from country to country.

A basic problem in attempting to prevent further introduction of pests and diseases into a country is that international trade and travel cannot be completely shut down. It is not even practical to totally prohibit the importation of plant material. Any quarantine action therefore must be to some extent a compromise with restrictions on trade and imports being commensurate with the risk of introduction and likely importance of a new pest or disease problem. If the risk and seriousness are great enough, as for instance with foot and mouth disease, severe measures are justified because of the drastic repercussions that would follow should the disease gain entry. Such measures might include a total trade ban (albeit temporary) on certain items and would certainly include total prohibition on import of live animals which could act as carriers. On the other hand the risk posed by a minor pest of a little grown ornamental plant could not justify such drastic measures.

Another problem in drawing up quarantine regulations is that plant and animal pests (and diseases) may be able to survive for some time away from their hosts. Quarantine restrictions therefore must also take into account likely non-host sources of infestation.

Several types of measures may be employed in quarantine programmes:

(a) *Total prohibition on imports of certain materials* — This can only be justified where a total ban imposes no great inconvenience and/or where the pest or disease is potentially so serious that risks cannot be taken (as with foot and mouth disease example quoted above). Soil is a prohibited import for many countries.

(b) *Goods provided with a Certificate of Health* — In the case of plant material, a Certificate of Health issued by authorities in the originating country may be required before entry is permitted. Unfortunately certificates from some countries are not reliable and other measures need to be taken.

(c) *Importation by licensed organisations only* — Many kinds of fresh fruit cannot be taken into many countries by individual travellers because of the risk of introducing fruit flies (Tephritidae). However, imports may be permitted by certain licensed importers under strictly controlled conditions.

(d) *Importation from certain designated areas only* — Where the overseas distribution of a pest or disease is limited, imports of plant material may be allowed from some countries or states but not from others.

(e) *Preferred types of plant material* — Seeds are less likely to harbour pests than vegetative parts of plants and thus, where a choice is possible, these are the preferred type of plant material for importation. If propagative material such as cuttings or bulbs is to be introduced greater care must be taken and more restrictions enforced to ensure that it is pest (and disease) free.

(f) *Chemical treatment* — Fumigation or other chemical treatment of plants or plant material, either before or immediately after entry, may be undertaken as an added precaution.

(g) *Isolation and inspection after entry* — It may be required that imported plants be grown initially in isolation and be inspected before distribution is permitted. This may cover a period of twelve months or more. If diseased or infested plants are found they may have to be destroyed.

Anyone wishing to import plant material of any kind should apply to the Ministry of Agriculture, Fisheries and Food for information and the required permit. Regulations vary according to the type of plant material (species, seeds, cuttings, etc) and the use to which it is to be put (consumption as human or animal food, for planting, or for propagating).

To attempt to smuggle in seeds or other plant material in violation of the regulations is a highly irresponsible act and may carry severe penalties. The consequences for the country should a serious pest or disease gain entry far outweigh the possible private gains from such acts.

MISCELLANEOUS CONTROL PROCEDURES

In addition to the main methods of dealing with pests that have now been discussed a number of other possibilities exist which do not fit into any of the described categories. These are, for the most part, rather specialised in nature and of limited application. Three are considered briefly:

(a) **Mechanical control**

Any means of mechanically trapping or killing insects, or the provision of barriers to prevent insects gaining access to plants or other materials, may be referred to as mechanical control. For example, an old recommendation to help control codling moth was to tie bands of sacking round the trunks of apple trees. The idea was that overwintering larvae would spin their cocoons in the bands which could then be removed and burned. Another example is provided by a device used in some flour mills to kill insects in the milling process. It consists of rapidly rotating studded steel plates between which the flour is fed. Mechanical force effectively destroys all stages of insects, including eggs.

Insect proof packaging of foods is an effective method of preventing infestation, provided of course that the material is free of pests to start with. Mechanical means of pest control are obviously of limited application but can be highly effective in appropriate circumstances.

(b) Physical control

This term refers primarily to situations where some physical factor in the environment may be modified to prevent or minimise a pest problem. For instance refrigerated air may be blown through stored grain to maintain the temperature too low for pests to multiply. Physical factors may also be used sometimes to attract insects to traps. A prime example is provided by light traps for night flying insects. Some pests are attracted and captured in enormous numbers in this way. However, the use of light traps in attempts to control pests has rarely been effective except under certain specialised conditions.

(c) The sterility principle in pest control

It perhaps seems odd at first sight that, given the opportunity, one might opt to sterilise a pest rather than kill it outright. Sterilising an insect does not prevent that individual inflicting injury but it does destroy the insect's reproductive capacity as effectively as if it were killed, and in addition these are positive advantages. Firstly, a sterile, but otherwise active, insect can seek out and neutralise the reproductive capacity of other individuals of the same species by mating with them. Secondly, if sterile insects are reared and released the effect is absolutely specific to the species concerned, which is a very desirable feature.

The idea of using sterility in pest control was first proposed early this century but it only became a practical possibility about 30 years ago with the availability of cheap sources of radiation from radioisotopes. These are two main approaches possible:

(1) to sterilise wild populations of a pest; or

(2) to rear and release large numbers of sterile individuals so that the wild population is outnumbered and suppressed.

The only practical means of sterilising wild populations of insects is by chemical sterilants. All those discovered so far, as earlier indicated, are highly hazardous to humans and so cannot be considered for other than experimental use. There is also the difficulty of trying to use them in such a way that only the pest species is affected and not other beneficial insects, a problem common to chemical control with insecticides. The most likely solution to the latter difficulty would be by development of bait formulations incorporating selective attractants.

The second approach, of rearing and releasing large numbers of sterile insects, enables other means of sterilisation to be used such as irradiation and various forms of genetic incompatibility. The principle behind this approach is that released insects, if sufficiently numerous, will compete with and reduce the reproductive capacity of the wild population. If the ratio of released sterile to wild fertile insects is sufficiently high the species may be eradicated. The practical feasibility of the technique was demonstrated with spectacular success in the 1950s and 1960s when the screw worm fly (*Cochliomyia hominivorax*), a serious pest of livestock, was eradicated in North America, firstly from the island of Curacao, then from Florida and later from the South-Western United States. In the latter instance, although complete elimination was initially achieved, control has not been entirely permanent as some reinvasion from Mexico has occurred.

This success caught the imagination of entomologists around the world and a great deal of effort was subsequently expended in attempts to apply the technique to other pests. Some limited successes were achieved but there were many failures. It is now realised that the sterile insect technique can only succeed if certain essential conditions are met. Cases of failure invariably involve the inability to ensure compliance with all the specified conditions. The most important of these are:

(1) Practical methods of mass rearing the insect must be available or capable of being developed.
(2) Sterilisation (normally with radiation) must be possible without seriously affecting vigour and mating competitiveness of the species.
(3) Released insects must not constitute a problem.
(4) Released insects must disperse adequately and seek out and mate with the wild population (males only of screw worm were initially used but later both sexes were released).
(5) The wild population must be outnumbered by a high ratio of released sterile insects (and this must be sustained for a sufficient period of time) if the population is to be suppressed.

The sterile insect technique is expensive but high initial cost could be justified if total eradication were feasible and the pest was a serious one (as with screw worm). However, if sustained releases are to be considered for limited term control (as may sometimes be the case), costs must be comparable to more conventional control methods.

It will be apparent therefore that the sterility approach to pest control is a very specialised one that is only applicable to certain pests and certain situations.

THE ECONOMICS OF PEST CONTROL

In considering the profitability or otherwise of pest control operations we may state in general terms that the cost of control measures for any pest should be at least balanced by the increased return resulting from control. This may be obvious and self evident but in practice there are many difficulties in expressing such a statement in quantitative terms.

The simplest situation is that of a cash crop. In such cases the value per unit of yield should be known fairly precisely, but unfortunately we often have only a very incomplete picture of the relationship between pest infestation level and crop loss for particular pests (see Chapter 8). This means that the benefit to be gained from controlling different pest populations cannot be accurately gauged. A second problem is that costs of control measures are sometimes difficult or impossible to establish. There is usually no difficulty with chemical control as costs of the chemical and costs of application should be known quite accurately. However, if biological control is involved, or cultural control measures of some sort, costing may be much more difficult. In the case of biological control by introduced parasites for example, there may be no direct cost to the grower or farmer at all but the research required has to be paid for indirectly by taxation over the wider community.

Despite these problems it is worthwhile to depict graphically the relationship between cost of control measures and profitability for an ideal situation where:

(a) we are concerned with a cash crop of known value per unit of yield;
(b) the relationship between injury and yield for a particular pest is well understood; and
(c) the cost of control measures is accurately known.

Such a relationship is presented graphically in Fig 28. It will be seen that the curve relating yield to pest injury is identical to that of Fig 21(a) which was for a pest attacking a non-yield forming part of the plant (and thus with a fairly high damage threshold level). Consider initially the situation where control measures are applied against a pest population exactly at the damage threshold level. The result would be no increase in yield and so there would be no return for expenditure incurred. At a slightly higher population level (pest injury level in the graph) some increase in yield would occur and therefore some increase in value of the crop, but still not enough to cover expenditure. With even higher pest (injury) levels the point is eventually reached where the cost of control measures is exactly balanced by the added value from increased yield. This is the **economic injury level**. At all points above the economic injury level the added value from increased yield will be more than the cost of control and thus the operation will be profitable.

Expressed another way we can say that control of pest populations above the economic injury level results in economic control

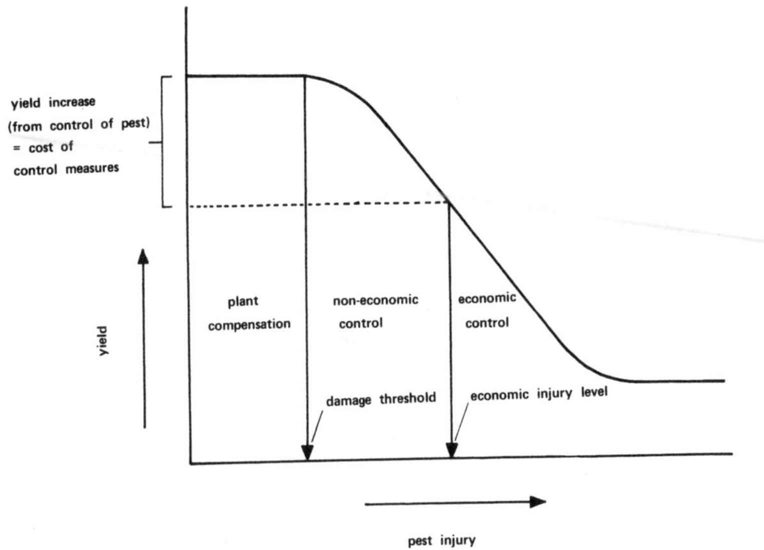

Fig. 28. The economics of pest control (for explanation see text).

because expenditure on control measures is exceeded by increased return resulting from control of the pest. In contrast, control applied against population levels below the economic injury level will be non-economic because the cost of the control measures will be greater than the added value of the crop. Ideally, decision making in pest control would be based on this sort of information. In practice however one usually has to operate on a much more rule of thumb basis because reliable quantitative data is inadequate.

Many pest control situations are in fact more complex and more difficult to quantify than that just described. How for instance does one put a monetary value on benefit gained from control of pests of ornamental plants, or of insect vectors of human disease? In such cases subjective value judgements are involved and it is not appropriate to try and establish profit or loss in purely monetary terms.

A further practical difficulty is that control treatments often have to be initiated *before* pest populations rise to damaging levels if they are to be effective. However, the prediction of changes in insect populations involves much uncertainty so that for valuable crops there is often a tendency to make treatments against pests (applications of insecticides in particular) whether they are really required or not. Such treatments can be regarded as insurance on the part of the grower to protect a valuable asset (the crop) but are undesirable because in many cases they involve unnecessary expenditure and excessive use of insecticides. The situation will only improve when our ability to predict pest population changes becomes more reliable.

In practice therefore decision making in pest control is often on a much more arbitrary basis than we would wish. One of the objectives of pest management (discussed in the next Chapter) is to improve the situation with better decision making models.

THE ADVANTAGES AND DISADVANTAGES OF DIFFERENT PEST CONTROL PROCEDURES

It should be apparent from the foregoing discussion of the major approaches to pest control that each has both advantages and disadvantages. Most of these have been considered under appropriate sections of the Chapter. They are summarised in Table 19.

The selection of control measures for a particular pest and situation depends to a large extent on the relative importance of these advantages and disadvantages which vary according to the circumstances. In fruit growing for example a very high level of pest control is of prime importance and highly toxic insecticides may be used despite their drawbacks. For household insect problems on the other hand, safety of any measures used must be the first consideration.

Some control measures tend to be antagonistic to or incompatible with others. This applies particularly to chemical control with insecticides and biological control with insect parasites and predators. An insecticide may be very effective against a pest but at the same time may kill large numbers of its parasites. The result is that good initial control is followed by rapid resurgence of the pest because it is no longer limited by its natural enemies. Such antagonistic reactions are not confined to insecticides. Apparently innocuous measures, such as various cultural control practices, may in fact be harmful to parasites and predators and should be carefully evaluated for such effects before they are adopted. Burning of crop residues for example may effectively limit the overwintering stages of a pest but may at the same time destroy its parasites.

A limitation of biological control is that it is often not sufficiently effective. One reason is that natural enemies take time to build up and during this time considerable crop damage may occur. In other cases parasites or predators of the pest in question may never reach numbers sufficient to suppress it to acceptable levels. Similarly, plant resistance to pests rarely provides more than partial control.

If means could be found of using two or more methods of control simultaneously so that they work together rather than against one another some of these limitations and disadvantages could be overcome. This is a major objective of pest management which forms the subject of the next Chapter.

Table 19. Summary of advantages and disadvantages of the major pest control methods

Method	*Advantages*	*Disadvantages*
Cultural control	Low cost (in most cases). Effects on non-target organisms minimal. No toxicity or residue problems.	Not always applicable. May not be sufficiently effective. Usually preventative in nature so that it requires detailed forward planning. May interfere with normal cultural operations.
Plant resistance	Low cost. No harmful effects on natural enemies of pests or other non-target organisms. No toxicity or residual problems. Normally permanent (but may break down in the face of new pest biotypes).	Pest resistant varieties or species of cultivated plants not available for all pests. Level of control may not be sufficient. Discovery and development slow. Resistant varieties may not be agronomically acceptable. Always preventative in nature, and thus requires forward planning.
Biological control (with parasites and predators)	Low cost. Permanent. (Apply to inoculation but not necessarily to inundation techniques.) Not harmful to non-target organisms. No toxicity or residue problems.	Not always applicable. Level of control may not be sufficient. Research costs are high and may not produce results.
Biological control (with pathogens)	Not harmful to other natural enemies or to other non-target organisms. No toxicity or residue problems.	Not always applicable. Control not usually permanent so there is recurring cost. May be difficult or expensive to produce.
Chemical control (with insecticides)	Applicable to most pests. Curative in effect. Farmer can apply when and where required. Enables high levels of control of most pests to be achieved and hence production of high quality produce.	Non-selective (may harm natural enemies and other non-target organisms). Resistance often develops. Often toxic to users and may present residue problems. Costs high and recurring as control is not permanent.

SELECTED REFERENCES

Anon. , 1969. *Insect-pest management and control.* National Academy of Sciences, Washington. 508 pp.

Anon. 1983. *Agricultural chemicals approval scheme: Approved products for farmers and growers.* Ministry of Agriculture, Fisheries and Food, HMSO, London. 246 pp. (revised annually)

Beroza, M. ed. 1976. *Pest management with insect sex attractants and other behaviour-controlling chemicals.* American Chemical Society, Washington. 192 pp.

Brown, A. W. A.; Pal, R. 1971. *Insecticide resistance in arthropods.* World Health Organization, Geneva. 491 pp.

Burges, H. D.; Hussey, N. W. 1971. *Microbial control of insects and mites.* Academic Press, London. 861 pp.

Davidson, G. 1974. *Genetic control of insect pests.* Academic Press, London. 158 pp.

De Bach, P. 1974. *Biological control by natural enemies.* Cambridge University Press, Cambridge. 323 pp.

Ebbels, D. L.; King, J. E. eds. 1979, *Plant health: the scientific basis for administrative control of plant diseases and pests.* Blackwell Scientific, Oxford. 334 pp.

Ennis, W. B. ed. 1979. *Introduction to crop protection.* American Society of Agronomy, Madison. 524 pp.

Gunn, D. L.; Stevens, J. G. R. eds. 1977. *Pesticides and human welfare.* Oxford University Press, Oxford. 278 pp.

Hedin, P. A. ed. 1977. *Host plant resistance to pests.* American Chemical Society. Symposium No 62. Washington. 286 pp.

Hewitt, W. B. ed. 1977. *Plant health and quarantine in international transfer of genetic resources.* CRC Press, Cleveland. 360 pp.

Huffaker, C. B.; Messenger, P. S. 1976. *Theory and practice of biological control.* Academic Press, New York. 788 pp.

Jacobson, M. 1965. *Insect sex attractants.* Interscience, New York. 154 pp.

MacBain Cameron, J. W. 1963. Factors affecting the use of microbial pathogens in insect control. *Annual Review of Entomology.* 8. 265-286.

McEwen, F. L.; Stephenson, G. R. 1979. *The use and significance of pesticides in the environment.* Wiley, New York. 538 pp.

Martin, H.; Worthing, C. eds. 1979. *Pesticide manual.* 6th edition. British Crop Protection Council, Croydon. 655 pp.

Matthews, G. A. 1979. Paperback edition, 1982. *Pesticide application methods.* Longmans, London and New York. 336 pp.

Maxwell, F. D.; Jennings, P. R. eds. 1980. *Breeding plants resistant to insects.* Wiley, New York. 683 pp.

Metcalf, R. L.; McKelvey, J. eds. 1976. *The future for insecticides: needs and prospects.* Wiley, New York. 524 pp.

Ordish, R. 1976. *The constant pest: a short history of pests and their control.* Davies, London. 240 pp.

Painter, R. H. 1951. (Paperback 1968). *Insect resistance in crop plants.* University Press of Kansas. 520 pp.

Roberts, D. A. 1978. *Fundamentals of plant-pest control.* Freeman, San Francisco. 242 pp.

Scopes, N. E. A. ed. 1979. *Pest and disease control handbook.* British Crop Protection Council, Croydon. 450 pp.

Shorey, H. H.; Gaston, L. K.; Jefferson, R. H. 1968. Insect sex pheromones. *Advances in Pest Control Research.* 8. 57-126.

Stern, V. M. 1974. Economic thresholds. *Annual Review of Entomology.* 18: 259-280.

Van den Bosch, R.; Messenger, P. S.; Gutierrez, A. P. 1982. *An introduction to biological control.* Plenum Press, New York. 247 pp.

Chapter 12

The Pest Management Concept

Pest management is fundamentally different from other approaches to pest control in that it aims (a) to utilise two or more control techniques together in an integrated fashion, (b) to make maximum use of natural mortality factors, and (c) to apply specific control measures only as and where necessary. An essential feature of pest management is its flexibility with decisions as to course of action (such as whether to apply an insecticide or not) based on the field situation.

Although many definitions of pest management have been proposed it is difficult to define briefly because of its inherent complexity. Perhaps the most suitable definition is that quoted by Brader (1979) — "pest management [is a] system that, in the context of the associated environment and the population dynamics of the pest species, utilises all suitable techniques and methods in as compatible a manner as possible and maintains pest populations at levels below those causing economic injury."

In this Chapter the essential components of pest management are first listed and each then considered in detail. To provide a concrete example reference is made to management of fruit tree red spider mite*. First however, some terms related to pest management must be discussed.

* — Although not yet practised to any extent in Great Britain management of fruit tree red spider mite (and some other mite species) by the procedures described in this Chapter is operational in several parts of the world including Australia, New Zealand, North America and Holland. More complex systems of integrated control have been developed for pests of some glasshouse crops (chrysanthemums in particular) which involve the use of parasites against leaf miner, predatory mites for control of spider mites. *Verticillium lecanii* for aphids, *Bacillus thuringiensis* for caterpillars and carbaryl for thrips. However these systems are not regulated by action thresholds based on monitoring and so have not been selected to illustrate this Chapter.

Terminology

A term that pre-dates pest management is *integrated control* which was applied originally to the harmonious integration of chemical and biological control. Most pest management programmes depend heavily on the successful blending of chemical and biological control methods and for this reason integrated control is now usually regarded as synonymous with pest management. However, pest management often involves considerably more than the mere joint use of chemical and biological methods, and is now generally the preferred term.

IPM is another term (or rather abbreviation) that occurs in some texts. It stands for "integrated pest management" and is no different in meaning from the shorter term pest management which is used throughout this book. IPM in the broadest sense embraces *all* "pests" and thus includes plant diseases and weeds as well as insects and other animal pests.

Pest management can be conveniently considered as part of resource management and there is thus a strong parallel with fish and game management. The latter both involve the management of biological systems in such a way as to provide maximum sustainable yield of fish or game. To achieve this, good understanding of the biological system concerned is essential. Successful manipulation of pests is similarly dependent on adequate understanding of the biological systems of which they are part. The objective though is to maintain a low "yield" of pest species rather than the reverse.

The essential components of pest management

Five major components can be identified in all pest management programmes:

(a) An understanding of the factors that regulate pest numbers (ie, of pest populations dynamics).
(b) The determination of pest damage thresholds and of economic thresholds.
(c) Means of monitoring populations of pests and their natural enemies.
(d) A decision making framework to determine action to be taken.
(e) Methods of selectively manipulating pest populations.

Each of these aspects is discussed below.

(a) *Understanding of the factors that regulate pest numbers*

The various factors that are involved in the natural regulation of insect numbers were discussed at length in Chapter 10. There it was pointed out that two factors of major importance for insects are natural enemies and the influence of climatic fluctuations. In attempting to develop a pest management programme attention must be paid to identifying important regulatory (key) factors and to developing as complete a picture as possible of the pest's population dynamics. So far our understanding of these matters is not as complete as we would wish but for a number of important pests present knowledge is sufficient for practical purposes.

The objective of these studies into pest population dynamics is several fold. First, with adequate understanding we should be able to predict changes in the populations of pests and thus enable pest outbreaks to be forecast. This in turn provides information as to the necessity or otherwise of implementing control measures. Secondly, an understanding of the factors that regulate the numbers of a pest may suggest ways of manipulating it. This may involve the importation of additional natural enemies from overseas, an aspect that is particularly important in some countries. Thirdly, an understanding of population dynamics may enable important events in the life cycle of pests to be predicted (such as emergence from eggs or pupae) so that control measures can be precisely timed. The latter is becoming an increasingly important aspect of pest management programmes.

In the case of fruit tree red spider mite it is known that predators are extremely important in regulating its numbers, so much so that their removal by injudicious use of toxic insecticides leads to damaging mite outbreaks. The most important predators are phytoseiid (typhlodromid) mites, particularly *Typhlodromus pyri*. The ladybird *Stethorus* sp. also feeds on fruit tree red spider mite and some other predatory insects are also involved. From information on population levels of fruit tree red spider mite throughout the growing season, and the ratio of predators present, changes in red mite numbers can be predicted and the necessity or otherwise for spraying determined (see under (d) below).

(b) *Determination of pest damage thresholds and of economic thresholds*

An essential feature of pest management is that pest populations are manipulated to maintain them just below damaging (threshold) levels (see Chapter 8 for discussion of damage thresholds). There are two reasons for this. One is that low populations of a pest will sustain populations of natural enemies whereas attempted complete elimination of a pest results also in severe reduction of its natural enemies. The second reason is that application of control measures only when the pest population is expected to exceed the damage threshold is much more economical than applications on a scheduled basis.

Ideally, one would like to ascertain the relationship between pest density and yield from zero to the highest levels encountered and at different times throughout the growing season (see Chapter 8), but this has been achieved for few pests. The minimum requirement is to determine the point at which yield just starts to be affected (the damage theshold).

Fruit tree red spider mite is an indirect pest in that it feeds on the leaves of apple trees rather than directly on the fruit. A moderate population of about 10 mites per leaf can cause leaf bronzing but the relationship between mite numbers and fruit yields is very complex, depending in part on the pattern of mite population build up in relation to seasonal growth of the crop.

A further complication in setting damage threshold levels for pests is that some cultivars of plants are more tolerant of pest injury than others and damage thresholds thus vary with the cultivar concerned. As a general principle in pest management the higher the damage threshold the better because natural enemies are more likely to be able to limit pest populations to moderate levels rather than to very low ones. Also many resistant plants limit the rate of growth of pests enabling natural enemies to exert more effective control. The introduction of pest resistant cultivars as a component of pest management is therefore very desirable.

In practice control measures almost invariably have to be initiated *before* pest populations reach the damage threshold level to prevent crop loss. Such a lower population level which signals the necessity for control action is referred to as the **economic threshold (action threshold** of some authors). Establishment of suitable economic thresholds is absolutely essential to the functioning of any pest management programme and those used for fruit tree red spider mite are described under (d) below.

(c) *Monitoring populations of pests and natural enemies*

Pest populations must be regularly monitored in pest management programmes in order to decide when to apply control measures. Where certain factors are known to be important in affecting pest numbers these also must be monitored. These factors may be physical, such as weather fluctuations, or biological, such as the incidence of natural enemies. Physical factors can be readily monitored by suitable instrumentation but keeping track of pests and their natural enemies requires special sampling procedures.

Populations of fruit tree red spider mite and its predators can be monitored on apple trees by systematic sampling and examination of leaves. Results are expressed as mean numbers of active stages of fruit tree red spider mite and its predators per leaf. Because of their short life cycles rather frequent sampling (at 1-2 week intervals) is necessary for this pest and its predators.

Sticky traps baited with sex attractant chemicals are being increasingly used for sampling the adult stages of flying insects such as codling moth.

Monitoring systems for pests and their natural enemies must be designed to suit the pest in question and must be practical in terms of time and labour involved. Cost of monitoring is an important factor in any pest management programme. Many such sampling systems still have to be developed or improved.

(d) *A decision making framework to determine action to be taken*

As pest management is essentially flexible in nature criteria for decision making as to action to be taken (such as whether to apply an insecticide or not) are of crucial importance. Some element of prediction of events is always involved in the decision making process. For example, monitoring of a field crop may indicate a moderate infestation of aphids but with a high level of parasitism. The prediction in this case may be that further significant increase in the aphid population will not take place and therefore that spraying is not necessary. Good reliability in predicting the course of events is clearly essential if pest management is to receive practical acceptance and for this reason some margin of safety is usually built into economic thresholds. Such a safety margin is also necessary to cover site to site and season to season variation that invariably occurs.

A basis for decision making has been developed for fruit tree red spider mite which takes into account population levels of the mite and relative numbers of predatory mites present. The criteria vary somewhat according to time of year. In early spring a miticide application is recommended when fruit tree red spider mite numbers reach three or more active stages per leaf if predatory mites are absent. In the presence of predatory mites spraying is not necessary until spider mite populations average five per leaf. In summer treatment in the absence of predators is recommended when the fruit tree red spider mite population averages two per leaf. If predatory mites are present and there are 2-5 active stages of fruit tree red spider mite per leaf spraying is only necessary if the ratio of predatory to spider mites is less than 1 to 7. If there are more than 5 spider mites per leaf the ratio of predators needs to be 1 to 4 if spraying is to be avoided.

(e) *Methods of selectively manipulating pest populations*

All pest management programmes attempt to make maximum use of natural controls (particularly control provided by parasites and predators), to apply artificial measures to supplement them only as necessary, and to achieve optimum timing of control measures. The most important artificial controls are insecticides but, as these can be highly toxic to natural enemies of pests, great care must be exercised in their use. Ideally, insecticides for use in pest management programmes should be toxic to the pest to be controlled but harmless to its natural enemies. Few insecticides so far discovered approach such a degree of selectivity but some are more selective than others. Cyhexatin for example is selective against fruit tree red spider mite and does not severely harm predatory mites. It is therefore one of the miticides recommended in pest management programmes for fruit tree red spider mite. In contrast, some other miticides are just as toxic to predatory mites as to phytophagous ones and therefore should not be used.

In apple orchards other important pests, such as codling moth and leafrollers, have to be controlled besides fruit tree red spider mite and at present insecticides are used for this purpose. At present azinphos-methyl is the main insecticide employed. This can be toxic to predatory mites but quite fortuitously strains of *Typhlodromus pyri* resistant to azinphos-methyl have appeared. This is a case where development of resistance to a chemical (by a predator) is beneficial as opposed to insecticide resistance in pest species. Azinphos-methyl resistant strains of this and other predatory mites are being artificially introduced into orchards where they do not occur naturally, and provide an important element in management of fruit tree red spider mite. Fungicides applied to fruit trees for control of such diseases as scab and mildew can also be harmful to predatory mites and careful choice of these is important, as with insecticides. Advisory leaflets concerned with management of orchard mites list pesticides recommended for use in such programmes and those that should be avoided.

Besides choice of insecticide there are some other possible ways of improving selectivity. One is to use insecticides at much lower than normal rates of use (a half, a quarter or even less). This may result in slightly less effective action against the pest but may be more than compensated by much better survival of natural enemies. Another possibility is to avoid overall treatment of the crop with insecticides. Instead, with each application, alternate rows (of fruit trees) or alternate swath widths (of field crops) are treated. The objective is to provide refuges relatively uncontaminated with insecticide in which parasites and predators can survive. With the next application the previously untreated strips are sprayed. Spot spraying of heavily infested areas only is another possibility.

A further possible technique for improving the selective action of insecticides is to use them in conjunction with specific chemical attractants for particular pest species. The latter approach has great potential with the continued discovery and improved availability of sex attractants for many insects of economic importance.

Discussion and conclusions

Pest management is much more complex than regular calendar application of insecticides and requires considerably greater expertise for its operation. The question may therefore be asked as to why it has been necessary to develop such programmes.

Several factors may be identified. In the first instance intensive use of broad spectrum insecticides has often led to the appearance of secondary pests and to the resurgence of primary pests due to harmful insecticidal effects on natural elements. Spider mites are a prime example. Secondly, from small beginnings, resistance of pests to insecticides has now developed to such an extent that it threatens continued effective chemical control of increasing numbers of pests around the world. Thirdly, and of more recent significance, is the rapid escalation in costs of pesticides, which is beginning to render their continued intensive use uneconomic in some situations.

Fortunately, some research aimed at developing pest management programmes was initiated many years ago by a few far sighted scientists before these problems became severe but much remains to be done. The practical development of pest management requires much research input extending over several growing seasons, and many programmes are still in a formative stage of development. Pest management does not therefore offer an immediate panacea for the problems that beset chemical pest control.

One aspect of pest management that many growers find difficult to accept is the presence within their crops of sub-economic levels of pests. During the chemical era of pest control growers have been educated to the philosophy that "the only good bug is a dead bug". Pest management in contrast emphasises the fact that not only are sub-threshold levels of pests harmless but that their presence is essential to maintain natural enemies which aid control on a continuing basis. Perhaps the most difficult idea of all to accept is that it may be desirable deliberately to introduce a pest, together with its natural enemies, into a crop at an early stage of growth to establish a balanced system from the outset. The management programmes developed for two-spotted red spider mite and glasshouse whitefly under glass, involve such a procedure. Education of growers and farmers to such concepts is important in the practical implementation of pest management programmes. This task will be facilitated if the reliability and economy of pest management can be demonstrated.

Several practical benefits result from effective pest management. Insecticide applications are often reduced to half or even less compared to full chemical programmes. This not only results in considerable cost saving but also reduces residue problems. However, some use of insecticides (in a carefully controlled manner) is nearly always required so that their cost is not entirely eliminated. Costs of monitoring must be offset against savings resulting from reduced use of insecticides. These costs may be considerable but total expenditure is often much in favour of pest management. A further important advantage of pest management is that resistant strains of pests are less likely to develop compared to full chemical programmes because they are not subjected to such intense selection pressure from a single factor. Finally, because of its sound ecological basis and its pest monitoring component, pest management should prove to be a more reliable method of pest control than calendar application of insecticides. Such reliability is particularly important with high value crops involving much financial input.

It has been emphasised that pest management involves the continual collection and interpretation of much information. One modern development that can greatly aid the handling and processing of such information is the computer. The recent availability of small and relatively cheap computers now makes it possible to store and provide immediate access to detailed information resulting from day to day monitoring of field areas where pest management is practised. Furthermore, with adequate programming, computers can be a valuable aid in the decision making process by determining the outcome of alternative courses of action with respect to pest populations. The ultimate development in this respect is the provision of direct links from grower or pest management adviser to computer facilities. Such systems are now starting to come into use in the most advanced pest management programmes.

In conclusion, we can view the emergence of pest management as an indication that pest control is at last becoming a truly applied science firmly based on applied ecology, in contrast to the more empirical approaches to pest problems that have existed up to now.

SELECTED REFERENCES

Apple, J. L.; Smith, R. F. 1976. *Integrated pest management.* Plenum Press, New York. 200 pp.

Boethel, D. J.; Eikenbary, R. D. 1979. *Pest management programmes for deciduous tree fruits and nuts.* Plenum Press, New York. 265 pp.

Brader, L. 1979. Integrated pest control in the developing world. *Annual Review of Entomology*. 24. 225-254.

Croft, B. A.; Howes, J. L.; Welch, S. M. 1976. A computer-based extension pest management delivery system. *Environmental Entomology*. 5. 20-34.

De Lara, M. 1981. Development of biological methods of pest control in the United Kingdom glasshouse industry. *Proceedings of 1981 British Crop Protection Conference,* pp. 599–607.

Gould, H. J. 1971. Large-scale trials on an integrated control programme for cucumber pests on commercial nurseries. *Plant Pathology*. 20. 149-156.

Gould, H. J. 1977. Biological control of glasshouse whitefly and red spider mite on tomatoes and cucumbers in England and Wales, 1975–76. *Plant Pathology*. 26. 57–60.

Hall, R. A. 1982. Control of whitefly, *Trialeurodes vaporariorum* and cotton aphid, *Aphis gossypii* in glasshouse by two isolates of the fungus *Verticillium lecanii*. *Annals of Applied Biology*. 101. 1–11.

Huffaker, C. B. ed. 1980. *New technology of pest control.* Wiley, New York. 624 pp.

Metcalf, R. L.; Luckman, W. eds. 1982. *Introduction to insect pest management.* 2nd edition. Wiley, New York. 577 pp.

Minks, A. K.; Gruys, P. eds. 1980. *Integrated control of insect pests in the Netherlands.* Centre for Agricultural Publishing and Documentation, Wageningen. 304 pp.

Rabb, R. L.; Guthrie, F. E. eds. 1979. *Concepts of pest management.* North Carolina State University, Raleigh. 242 pp.

Stern, V. M. 1974. Economic thresholds. *Annual Review of Entomology*. 18. 259-280.

Watson, T. F.; Moore, L.; Ware, G. W. 1975. *Practical insect pest management.* W. H. Freeman, San Francisco. 196 pp.

Chapter 13

Information Required in Dealing with a Pest Problem

Although pest problems often have to be tackled out of sheer necessity from a quite inadequate basis of information, this is not a satisfactory situation. Wherever possible essential background information should be gathered and evaluated before control measures are initiated. What constitutes such information has been considered at many places throughout this book. The main elements are summarised here for rapid reference. They are listed in the order that should be followed in practice.

1. Identify the pest

This is a very important first step because the name provides the key to all published information about the pest, including the most appropriate control measures. Sometimes identification to group level only (such as aphids) is adequate, but usually it is preferable to identify the pest to species because there may be important differences between closely related insects. Symptoms of plant injury can be a valuable aid to identification as well as features of the pest itself.

Correct identification is essential. Mis-identification of a pest can only lead to wrong information which is worse than none at all. If the identity of a pest is in doubt the services of a specialist should be sought rather than a guess made. Specimens and pest damaged plant material may be sent to a recognised laboratory such as the Agricultural Development and Advisory Service regional centres of the Ministry of Agriculture, Fisheries and Food.

2. Obtain information about:

(a) *Life cycle*

It is important to know the sequence of developmental stages (from egg to adult) and their duration, how many generations occur per year and the method of overwintering, and whether the pest reproduces parthenogenetically (without mating) or is viviparous (gives birth to living young).

In addition to understanding these features of the life cycle, ability to recognise early developmental stages is often essential for pests which require early initiation of control measures for effective control.

(b) *Habits*

This covers a great many aspects but important features are stages in the life cycle which cause plant injury, type of mouthparts and mode of feeding, part of the plant attacked, and whether the insect feeds externally or tunnels into the plant. It is also important to know whether the pest may be involved in transmission of disease organisms in addition to any direct injury caused.

(c) *Plant host range*

Knowledge of the plant host range of pests is important because it indicates what other plants may be at risk (in addition to that on which the pest was discovered) and whether non-cultivated plants (weeds, hedgerow plants) can act as reservoirs of infestation. Plant host range is also important in relation to what crops can safely follow that on which the pest initially occurred.

(d) *Natural controlling factors*

Some pests are held in check most of the time by natural regulating factors of which parasites, predators and disease organisms are the most important. It is very desirable to know what the significance of these natural enemies is in particular circumstances as this may affect the choice of control measures.

(e) *Mobility and capacity for reinfestation*

Some pests are highly mobile and flying adults may be able to rapidly re-invade an area following local elimination. Other species in contrast have very limited ability to move and thus are very slow to reinfest. It is obviously important to be aware of a pest's capabilities in this respect.

3. Decide the need for control measures

The mere presence of a pest species at low population densities does not justify the application of control measures unless the population is above the damage threshold (see Chapter 8) or likely to rise above it if left unchecked. However, some pests do require treatment while populations are still insignificant to prevent them rising to damaging levels later.

Cost of control measures is a major factor that must also be taken into consideration. Expenditure on control measures can only be justified when the cost is at least balanced by increased value of the crop resulting from control of the pest; that is, when the pest population exceeds the economic injury level (see Chapter 11). Control measures should therefore only be initiated when they are economically justifiable. Exact quantification is unfortunately often difficult.

4. Select control measures

All possible measures that may be useful against a pest should be considered before the most appropriate is selected. It is important to know whether parasites, predators or other factors help to keep the pest in check as this may influence choice of control measures.

Can cultural control measures assist?

Are resistant varieties or species of plants available?

Have pest management techniques for the pest been developed? Answers to such questions should be sought before a choice of control methods is made. It is most important that up to date information and advice on these matters be obtained. This is available for many pests in the form of advisory leaflets produced by the Ministry of Agriculture, Fisheries and Food, referred to in Chapter 6.

If insecticides must be used the following aspects need consideration in selecting the most appropriate for a particular purpose:

(a) effectiveness for the pest to be controlled (contact, systemic and fumigant properties may be important);
(b) toxicity;
(c) formulation and method of application most suited to the circumstances;
(d) timing of application in relation to stage of development of the pest and stage of growth of the crop;
(e) whether repeat application after an appropriate interval is necessary. Some insecticides kill one stage of an insect but not others; for example active stages but not eggs. A second application after a suitable period of time may then be necessary;
(f) selectivity for beneficial organisms, particularly parasites and predators of the pest to be controlled;
(g) cost.

EPILOGUE

It should be apparent to readers of this book that there is no easy path to pest suppression. Pests are living organisms which successfully exploit resources inadvertently provided by human activity and which respond (in various ways) to threats to their existence. Pest suppression thus requires continual vigilance.

To provide permanent long-term pest control we must learn to work with the forces of nature rather than against them. This is inherent in the pest management approach, and to achieve it effectively will require interdisciplinary team work and not just the services of entomologists.

Although pest control must always play a secondary role to that of production in agriculture and horticulture, it should be considered an integral part of the production process and not a separate specialty to be called in only when things go wrong.

GENERAL REFERENCES

Becker, P. 1974. *Pests of ornamental plants.* Bulletin No 97. 3rd Edition. HMSO, London. 175 pp.

Buczacki, S.; Harris, K. 1981. *Collins guide to the pests, diseases and disorders of garden plants.* Collins, London. 512 pp.

Edwards, C. A.; Heath, G. W. 1964. *The principles of agricultural entomology.* Chapman and Hall, London. 418 pp.

Hussey, N. W.; Read. W. H.; Hesling, J. J. 1969. *The pests of protected cultivation.* Arnold, London. 404 pp.

Johnson, W. T.; Lyon. H. H. 1972. *Insects that feed on trees and shrubs; an illustrated practical guide.* Cornell University Press, Ithaca. 464 pp.

Jones, F. G. W.; Jones, M. G. 1974. *Pests of field crops.* 2nd Edition. Arnold, London. 448 pp.

Massee, A. M. 1954. *The pests of fruit and hops.* 3rd Edition. Crosby and Lockwood, London. 284 pp.

Munro, J. W. 1966. *Pests of stored products.* Hutchinson, London. 234 pp.

Pirone, P. P. 1978. *Diseases and pests of ornamental plants.* 5th Edition. Wiley, New York. 566 pp.

APPENDIX 1. CATALOGUE OF INSECTICIDES AND ACARICIDES*

Common name (alternative name in brackets)	*Main trade name(s)*	*Chemical group*†	*Acute mammalian toxicity LD 50mg/kg (rat unless stated otherwise)* **oral**	**dermal**	*Main uses*	*Remarks*
aldicarb	Temik	C	0.6	2.5	aphids, leafminers, spider mites, nematodes	systemic, very highly hazardous, toxic to fish and wildlife; granular formulation only
aldrin	Aldrex	OC	40–60	>200	wireworms, vine weevil, narcissus flies	permitted uses restricted due to long residual life; harmful to fish
amitraz	Mitac	O	800	1,600	spider mites, suckers	for use on apple and pear only; harmful to fish
azinphos-methyl	Gusathion	OP	7–13	280	caterpillars, sawflies, weevils, aphids, spider mites	very highly harzardous, toxic to bees; available only in mixture with demeton-S-methyl sulphone
bendiocarb	Garvox	C	34–64	560–800	pygmy beetle, wireworms, millipedes, springtails and symphilids on sugar beet	highly hazardous, toxic to fish, stock and wildlife; granular formulation only
carbaryl	Sevin, Murvin	C	400	>500	caterpillars, earthworms in turf, fruit thinning	highly toxic to bees
carbofuran	Yaltox	C	8	120	cabbage root fly, onion fly, flea beetles, leaf miners, aphids, nematodes	systemic, very highly hazardous, toxic to bees, fish and wildlife; granular formulation only
carbo-phenothion	Trithion	OP	7–30	800	wheat bulb fly	used for seed treatment of autumn sown wheat only; toxic to fish and wildlife
chlorfenvin-phos	Birlane, Sapecron	OP	10–150	30–100	cabbage root fly, frit fly, carrot fly, mushroom flies	very highly hazardous, toxic to bees, fish and wildlife
chlorpyrifos	Dursban, Spannit	OP	130–160	>2,000	caterpillars, aphids, sawflies, frit fly, wheat bulb fly, spider mites	toxic to bees and to fish

chlorpyrifos-methyl	Reldan	OP	1,600–2,200	>2,000 (rabbit)	insects and mites in grain	toxic to fish
cyhexatin	Plictran	T	230–650	>2,000	spider mites	selective for predatory mites, toxic to fish
cypermethrin	Cyperkill	SP	120–770	>3,000	caterpillars, suckers, aphids	extremely toxic to fish, dangerous to bees
DDT	—	OC	300–500	2,500	chafer grubs, cutworms, leatherjackets	permitted uses restricted due to long residual life; toxic to bees, harmful to fish
deltamethrin	Decis	SP	120–500	>1,800 (rabbit)	caterpillars, suckers, aphids, scale insects, whitefly	toxic to bees and to fish
demeton-S-methyl	Duratox, Metasystox, Mifatox	OP	40	1,000	aphids, sawflies, spider mites	systemic; very highly hazardous, toxic to bees, fish and wildlife
derris (rotenone)	—	B	132–1,500	—	caterpillars, aphids, sawflies, spider mites	toxic to fish
diazinon	Basudin	OP	300–600	>1,200	cabbage root fly, carrot fly, mushroom flies	toxic to bees, fish and wildlife
dichlorvos (DDVP)	Nogos	OP	25–30	75–900	aphids, thrips, whitefly, mushroom flies, spider mites	highly hazardous, vapour action, short persistence, toxic to bees
dicofol	Kelthane	OC	550–2,000	1,000–1,200	spider mites	not selective for predatory mites

APPENDIX 1 *(continued)*

Common name (alternative name in brackets)	*Main trade name(s)*	*Chemical group*†	*Acute mammalian toxicity LD_{50}mg/kg (rat unless stated otherwise)* oral	dermal	*Main uses*	*Remarks*
diflubenzuron	Dimilin	O	c.4,600 (mouse)	>2,000 (rabbit)	caterpillars, sciarid larvae	acts by disrupting moulting; long residual life
dimethoate	Rogor	OP	200–300	700-1,150	aphids, suckers, sawflies, thrips, wheat, bulb fly, spider mites	systemic, highly hazardous, toxic to bees, fish and wildlife
disulfoton	Disyston, Solvigran	OP	4	50	aphids, carrot fly	systemic, very highly hazardous, toxic to bees, fish and wildlife; granular formulation only
DNOC	Sandolin Cresofin (in petroleum oil)	O	25–40		overwintering stages of aphids, sucker, scale insects	for dormant season use only; toxic to fish and wildlife
endosulfan	Thiodan	OC	35	75–680	tarsonemid and eriophyid mites	highly hazardous, toxic to bees and livestock, very toxic to fish
ethiofencarb	Croneton	C	400–500	>1,150	aphids in potatoes and sugar beet	systemic, toxic to bees and fish
etrimfos	Ekamet	OP	1,800	>2,000	caterpillars in brassicas, leatherjackets	toxic to bees and fish, harmful to livestock and wildlife
fenitrothion	Dicofen Fenstan	OP	250–670	1,500–3,000	aphids, suckers, caterpillars, sawflies, stored grain pests	toxic to bees, harmful to fish, livestock and wildlife
fonofos	Dyfonate	OP	8–18	c.150 (rabbit)	cabbage root fly, wheat bulb fly	highly hazardous, toxic to fish and wildlife, granular formulation only

formothion	Anthio	OP	360–500	>1,000	aphids	toxic to bees, harmful to fish and livestock
HCH (lindane)	Gamma-Col	OC	200	500–1,000	aphids, caterpillars, saw-flies, beetles, flies	long residual life, toxic to bees, harmful to fish and wildlife
heptenophos	Hostaquick	OP	96–120	c.2,900	aphids	systemic, short lived, toxic to bees and fish
iodofenphos	Elocril	OP	2,100	>2,000	cabbage root fly, caterpillars on cabbage and sprouts	toxic to fish
malathion	Malastan	OP	1,400–1,900	>4,000	numerous insects and spider mites	low toxicity to humans but toxic to bees and fish
mephosfolan	Cytro-lane	OP	9	10	aphids on hops only	systemic, very highly hazardous, soil application only, toxic to bees, fish and wildlife
methidathion	Supracide	OP	20–48	25–400	aphids on hops	highly hazardous, toxic to bees, fish, live-stock and wildlife
methiocarb	Draza	C	100–135	350–700	strawberry seed beetle, slugs and snails	toxic to fish, bait and pellet formulations only
methomyl	Lannate	C	27	>1,600	aphids on hops	systemic, very highly hazardous, toxic to bees, fish and livestock
mevinphos	Phosdrin	OP	3–5	90	aphids, caterpillars, beet leaf miner	systemic, short lived, very highly hazardous, toxic to bees, fish, livestock and wildlife
nicotine	—	B	50–60	50 (rabbit)	aphids, capsids, leaf miners, thrips	highly hazardous, toxic to bees, fish, livestock and wildlife
omethoate	Folimat	OP	50	1,400	aphids on hops and wheat bulb fly	systemic, highly hazardous, toxic to bees, fish, livestock and wildlife

APPENDIX 1 (*continued*)

Common name (alternative name in brackets)	*Main trade name(s)*	*Chemical group*†	*Acute mammalian toxicity LD_{50} mg/kg (rat unless stated otherwise)* oral	dermal	*Main uses*	*Remarks*
oxamyl	Vydate	C	5.4	37	pygmy beetle, leaf miners on beet and tomato, whitefly, spider mites	systemic, very highly hazardous, toxic to fish and wildlife, granular formulation only
oxydemeton-methyl	Metasystox R	OP	65–80	250	aphids, suckers, sawflies, spider mites	systemic, highly hazardous, toxic to bees, fish, livestock and wildlife
permethrin	Ambush	SP	1,500	>4,000	codling, tortrix, winter moths and other caterpillars, suckers	toxic to bees, very toxic to fish
petroleum oil	—	O	low toxicity		spider mites, mealybugs, scale insects	normally for use under glass
phorate	Terrathion	OP	2–3	70–300	aphids, frit fly, wireworms, capsids, leafhoppers	systemic, very highly hazardous, toxic to fish, livestock and wildlife, granular formulations only
phosalone	Zolone	OP	120–170	1,500	aphids, caterpillars, spider mites	toxic to bees, fish and livestock
pirimicarb	Aphox Pirimor	C	147	600	aphids	hazardous, selective aphicide, harmful to livestock
pirimiphos-methyl	Actellic Blex	OP	1,100	>2,000 (rabbit)	aphids, caterpillars, flies, stored grain pests	toxic to bees and fish
quinalphos	Savall	OP	60–140	800–1,400	caterpillars, leatherjackets	highly hazardous, toxic to bees, fish, livestock and wildlife
quinomethionate	Morestan	O	2,500–3,000	>500	spider mites	also has fungicidal action

tar oil	Mortegg	O	low toxicity but can cause dermatitis		overwintering stages of aphids, scale insects, suckers and moths	dangerous to fish
tetradifon	Tedion	OC	>1,400	>1,000 (rabbit)	spider mites	selective for predatory mites
thiofanox	Dacamox	C	8	39 (rabbit)	aphids on potatoes and sugar beet	systemic, very highly hazardous, toxic to fish and wildlife, granular formulations only
hiometon	Ekatin	OP	100	>200	aphids	systemic, highly hazardous, toxic to bees, fish, livestock and wildlife
riazophos	Hostathion	OP	80	>25 (rabbit)	caterpillars, flies, seed weevils, spider mites	highly hazardous, toxic to bees, fish, livestock and wildlife
richlorphon	Dipterex	OP	650	>2,800	cabbage root fly, caterpillars, beet leaf miner	toxic to fish
amidothion	Kilval	OP	64–100	1,160 (rabbit)	aphids, sucker, sawfly, spider mites	systemic, hazardous, toxic to bees, livestock and wildlife

— Active ingredients for which there are approved products in Great Britain (1983)

— C, carbamate
OC organochlorine
OP organophosphate
SP synthetic pyrethroid
T tin compound
B botanical insecticide (plant derived)
O other

APPENDIX 2. GLOSSARY

abdomen — the posterior of the three body regions of an insect.

Acari — the order of arthropods within the class Arachnida that comprises the mites and ticks.

acaricide — (alternative term, miticide) — chemical used for control of mites.

accessory gland — gland associated with the insect reproductive system but fulfilling some function secondary to the reproductive process. Present in both sexes of most insects.

acetylcholine — a chemical substance produced at nerve connections (synapses) each time a nerve impulse is transmitted.

action threshold — see economic threshold.

active ingredient (a.i.) — the biologically active chemical in pesticide preparations.

aedeagus — the male copulatory organ of an insect, usually taken to include accessory structures besides the penis.

aerosol — very finely divided particles (of insecticide solution) in the air; often applied to the pressurised canisters which generate an aerosol.

aestivate — to enter a state of dormancy over the summer period. cf. hibernate.

a.i. — see active ingredient.

air assisted sprayer — a sprayer in which an air stream is employed to help carry spray droplets to the object being sprayed.

air blast sprayer — see mist blower.

alate — possessing wings. cf. apterous.

Ametabola — (alternative term, Apterygota) — subclass of primitive wingless insects that do not undergo metamorphosis. Comprises the orders Thysanura, Diplura, Protura and Collembola. cf. Metabola.

Anoplura — an alternative name for the insect order Siphunculata which comprises the sucking lice.

antenna (pl. antennae) — sensory feelers located on the head of insects bearing minute organs of touch and smell.

anther — the male part of a flower which produces pollen.

antibiosis — a mechanism of plant resistance to pests whereby harmful effects occur to the insect through feeding on the plant.

aorta — the forepart of the dorsal blood vessel of insects. In contrast to the hindpart (heart), the aorta is without valves. It discharges into the head region.

Aphaniptera — an alternative name for the insect order Siphonaptera which comprises the fleas.

apiculture — the culture of bees, honeybees in particular.
apodeme — an internal projection of the exoskeleton of insects and serving for muscle attachment.
apodous — without legs.
appendage — an external limb or similar structure arising from segments of the head, thorax or abdomen.
apterous — without wings. cf. alate.
Apterygota — (alternative term, Ametabola) — an alternative term to Ametabola for the subclass of primitive insects that are wingless and do not undergo metamorphosis. cf. Pterygota.
Arachnida — a class of arthropods closely related to insects which includes spiders, mites, ticks and scorpions.
Araneae — the order of arthropods within the class Arachnida that comprises the spiders.
Arthropleona — a sub-order of springtails (Collembola) having a distinctly segmented body of elongate form. cf. Symphypleona.
Arthropoda — the phylum of animals to which insects and other groups belong and characterised by the possession of an external jointed skeleton, segmented body and jointed limbs.
atomise — to break up into fine droplets (as applied to spray liquids).
attractant — a chemical or physical source which induces insects to move towards it. cf. repellent.
augmentation — mass culture and local release of parasites, predators or pathogens to provide short term biological control of pest species.
bait — an insecticide formulation to which a pest is attracted and which is usually ingested by the pest.
behavioural resistance — a mechanism in insects of resistance to insecticides which involves a behavioural change whereby contact with the insecticide is avoided.
binomial system — the system of dual latin names for plants and animals devised by the Swedish naturalist Linnaeus and now almost universally applied to living organisms.
biological control — the use of living organisms to control pest species.
biomass — the weight of living organisms in any given habitat.
bionomics — the study of habits, life histories and adaptations of living organisms.
biotic potential — (alternative term, reproductive potential) — the maximum possible rate of increase of an organism in the absence of any limiting factors.
biotype — a genetically distinct strain or sub-group of a species

distinguished by some behavioural or physiological difference but indistinguishable morphologically. Usually applied to strains of insect species which can overcome resistant plants.

broad spectrum (as applied to insecticides) — having an effect on a wide range of insects.

bursa copulatrix — a specialised part of the female insect reproductive system into which sperm is received from the male during copulation.

campodeiform — a form of insect larva (or nymph) with well developed legs, prominent antennae and usually several tail filaments.

capitulum — a small head; in particular the false head of certain mites and ticks on which the mouthparts are borne.

carbamate (insecticide) — a group name applied to insecticides which are based on the carbamic acid nucleus. Carbaryl is an example.

carcinogenic — cancer inducing.

cardo — a basal sclerite forming part of the maxilla in the mouthparts of insects.

caste — structurally and functionally distinct group within the colony of a social insect, eg, worker caste.

cercus (pl. cerci) — one of a pair of appendages at the tip of the abdomen of insects.

chelicerae — the inner portions of the mouthparts of mites; modified to piercing stylets in plant feeding mites.

chemical control — the use of chemicals to kill, deter or in any way suppress pest populations.

chemoreception — sensory perception of chemical substances; includes both taste and smell.

chemosterilant — a chemical which has the property of sterilising an insect but which does not kill it.

Chilopoda — the order of arthropods within the class Myriapoda that comprises the centipedes.

chitin — a major constituent of the cuticle of insects; chemically a nitrogenous polysaccharide.

chlorinated hydrocarbon — see organochlorine.

chlorosis — a yellowing of the foliage of plants due to lack of chlorophyll.

cholinesterase — an enzyme which breaks down acetylcholine. The latter is formed at nerve connections (synapses) each time a nerve impulse is transmitted.

chronic toxicity — the toxicity of a pesticide (to higher animals) when administered in small sub-lethal doses over a long period of time.

chrysalis — the pupal stage of butterflies and moths.
cibarial pump — see cibarium.
cibarium — (alternative term, cibarial pump) — a suctorial pump present in the head region of insects with piercing and sucking mouthparts (Hemiptera).
claspers — a pair of processes at the end of the abdomen of male insects which serve to clasp the female during copulation.
class — a division of the animal kingdom below a phylum but above an order, eg, the class Insecta.
coarctate — pupal form in which the last larval skin is retained as an extra covering; occurs particularly in higher Diptera.
cocoon — a silken covering produced by some insects which encloses and protects the pupa.
Coleoptera — the order of insects comprising the beetles (including weevils).
Collembola — the order of insects comprising the springtails.
compatible (as applied to pesticides) — safe to mix together.
compatibility (as applied to pesticides) — refers to whether or not different pesticides can be mixed together for application. Safe mixtures are compatible. Unsafe mixtures are incompatible.
complete metamorphosis — metamorphosis in which the insect proceeds through four distinct developmental stages — egg, larva, pupa and adult. cf. incomplete metamorphosis.
contact insecticide — one which is able to kill insects by contact with the external body surface.
control (as supplied to pest species) — reduction or maintenance of pest populations below the damage threshold.
cornea — the outer transparent layer of the eye.
cornicle — a tubular process (paired) which occurs on the abdomen of aphids.
corpora allata — a pair of glands associated with the brain of insects which secrete juvenile hormone.
coverage — the degree to which pesticide applied reaches the area intended. See ground cover, plant cover, target cover.
coxa (pl. coxae) — the uppermost joint of insect legs by which they are attached to the thorax.
crawler — the first instar nymph of scale insects which possesses legs and thus enables dispersal to take place.
crochets — minute spines which form a pattern on the extremity of false legs of lepidopterous caterpillars.
crop (as applied to insect anatomy) — the distended part of the foregut of insects that receives ingested food from the oesophagus.
crop rotation — the practice of planting botanically unrelated crops

on the same land in successive years.

cross pollination — transfer of pollen between different plants of the same species.

Crustacea — a class of arthropods, mostly aquatic in habit, to which crabs and crayfish belong.

crystalline cone — the clear conical shaped structure lying immediately beneath the cornea in each unit (ommatidium) of insect compound eyes.

cultivar — (alternative term, variety) — cultivar variety of a plant.

cultural control — manipulation of cultural practices to provide control of a pest.

cuticle — the outermost non-cellular layer of an insect's body wall.

damage threshold — pest population above which crop loss occurs.

density dependent (factor) — a factor regulating populations of living organisms, whose influence varies with the population density of the organism concerned.

density independent (factor) — a factor regulating populations of living organisms, whose influence is independent of the population density of the organism concerned.

deposit — the amount of an insecticide on a surface immediately following application.

dermal toxicity — the toxicity of an insecticide (to higher animals) when applied to the intact skin.

Dermaptera — the order of insects comprising the earwigs.

deterrent — a chemical substance which deters feeding or oviposition of an insect. cf. stimulant.

detoxication — the process by which a poison is rendered harmless.

deutonymph — the third instar stage in the development of mites and ticks. cf. protonymph.

diapause — a state of suspended activity and metabolism which may occur at any stage in the life cycle of an insect as a means of surviving unfavourable conditions. Diapause is normally brought about by a change in environmental conditions (especially reduced daylength) and requires a particular stimulus for its termination.

Dictyoptera — the order of insects comprising the cockroaches and mantids.

dimorphism — a difference in bodily form, size or colour between two groups of individuals within a species.

diploid — having the chromosomes present in homologous pairs, ie, with a chromosome complement of 2n. cf. haploid.

Diplopoda — the order of arthropods within the class Myriapoda that comprises the millipedes.

Diplura — the order of insects that comprises the two-pronged bristle tails.

Diptera — the order of insects that comprises the true flies which are characterised by possession of only one pair of wings.
dorsal — the top or upper (back) surface of an animal. cf. ventral.
drone — the male of social bees and wasps.
dun — (alternative term, sub-imago) — winged but sexually immature mayfly adult which has to undergo another moult.
EC — see emulsifiable concentrate.
ecdysis — (alternative term, moulting) — the process of shedding the skin (exoskeleton) undertaken periodically by insects and other arthropods.
ecdysone — (alternative terms, moulting hormone, MH) — the hormone in insects which initiates the moulting process.
economic control — control of a pest such that expenditure on control measures is more than compensated by increased value of yield.
economic injury level — pest population level at which expenditure on control measures is just balanced by the value of increased yield resulting from control of the pest.
economic threshold —(alternative term, action threshold) — pest population level at which control measures should be initiated to prevent the population rising above the economic injury level.
ecosystem — the ecological system of a given area; composed of living plants and animals and non-living components.
ectoparasite — a parasite which lives externally on the body of its host. cf. endoparasite.
eelworm — see nematode.
ejaculatory duct — the duct in male insects which conveys sperm to the copulatory organ.
elytron (pl. elytra) — the hard horny forewings of beetles (Coleoptera) which act as protective cases for the membranous hind wings.
Embioptera — a small order of tropical insects which live gregariously in silken nests; sometimes referred to as web spinners.
emulsifiable concentrate (EC) — an insecticide formulation in which the active ingredient is dissolved in a non-aqueous solvent to which emulsifiers are added. An emulsion is formed when added to water.
endemic (with reference to living organisms) — naturally occurring within a region.
endoparasite — a parasite which lives internally within the body of its host. cf. ectoparasite.
Endopterygota — (alternative term, Holometabola) — the subdivision of insects that undergo complete metamorphosis

with larva, pupa and adult stages. cf. Exopterygota.

enzyme — a biochemical which promotes a specific chemical reaction.

Ephemeroptera — the order of insects comprising the mayflies.

eradication — complete elimination of a species from an area or country.

exarate (as applied to insect pupae) — having legs and other appendages separately encased (as in the pupae of beetles, bees and wasps).

Exopterygota — (alternative term, Hemimetabola) — the subdivision of insects that undergo incomplete metamorphosis. cf. Endopterygota.

exoskeleton — the external body wall of insects and other arthropods. Usually involves hardened plates over at least some of its area.

family — the main unit of classification into which genera are grouped. Insect family names always end in -idae, eg, Scarabaeidae.

fat body — the principal food storage organ within the insect body; usually of diffuse and variable form providing "packing" around many internal organs.

flabellum — the small spoon-shaped tip of the "tongue" of bees.

formulation — the form in which an insecticide is prepared for practical use.

frass — insect excreta.

fumigant — an insecticide which is sufficiently volatile to act at a distance in the vapour form.

furca — a forked structure; as for example the forked skeletal support in the upper part of the mouthparts of flies.

galea — the outer arm of the "pincers" present on the maxilla of insect mouthparts.

gall — a swelling or outgrowth produced by a plant as a result of attack by an insect, mite or micro-organism.

ganglion (pl. ganglia) — a distinct swelling of nervous tissue in which many interconnections between nerve endings occur.

gastric caeca — an outgrowth from the forepart of the mid-gut of insects.

genitalia — the external reproductive organs of insects (male or female).

genus (pl. genera) — the next highest category to species in the classification of organisms.

gizzard — the portion of the gut of insects immediately behind the crop; often thick-walled and muscular.

glossa — the inner part of the extremity of the labium of insect mouthparts.

gradual metamorphosis — alternative term for incomplete or partial metamorphosis.

granule — an insecticide formulation which consists of relatively large (1-3 mm) particles of an inert material (carrier) which are impregnated with insecticide or on which the insecticide is coated.

gregarious — the habit of living together in groups.

ground cover — the degree to which chemical applied covers the ground area being treated.

Grylloblattodea — a small order of primitive insects related to the cockroaches and crickets for which there is no common name.

haemocoel — the body cavity of insects filled with blood (haemolymph) within which the internal organs lie.

haemolymph — the unpigmented blood of insects.

haltere(s) — the knob-shaped modified hind wings of Diptera which act as balancing organs.

haploid — having a single set of unpaired chromosomes, ie, with a chromosome complement of 1n. cf. diploid.

"hard" pesticide — one which persists for a long time in the environment; particularly applies to organochlorine insecticides.

head capsule — the hardened external skeleton of the head of insects.

heart — the hindpart of the dorsal blood vessel of insects. The heart is muscular, equipped with inlet valves (ostia) and acts as a circulatory pump.

Hemimetabola — (alternative term, Exopterygota) — the subdivision of insects that undergo incomplete metamorphosis. cf. Holometabola.

Hemiptera — the order of insects that includes cicadas, leafhoppers, scale insects, aphids, plant bugs and many other groups; characterised by the possession of piercing and sucking mouthparts.

herbivor — any animal (including insects) which feeds on living plants.

hermaphrodite — having both male and female sexual organs in the same body.

Heteroptera — a sub-order of the insect order Hemiptera. The Heteroptera includes the true bugs. cf. Homoptera.

hibernate — to enter a state of dormancy over the winter period. cf. aestivate.

high volume spraying — the application of pesticides in dilute form to thoroughly wet the object being sprayed; usually involves the application of 1120 or more litres per hectare (100 gals per acre).

Holometabola — (alternative term, Endopterygota) — the subdivision of insects that undergo complete metamorphosis. cf. Hemimetabola.

Homoptera — a sub-order of the insect order Hemiptera. The Homoptera includes cicadas, leafhoppers, aphids and scale insects. cf. Heteroptera.

honeydew — excess sugary plant sap excreted from the hind end of the gut by sap sucking insects.

hormone — a chemical, secreted into the bloodstream, which produces a specific effect on some bodily organ or process.

host — a species of plant or animal on which an insect feeds.

host plant — a species of plant on which a particular insect species feeds.

Hymenoptera — the order of insects that includes sawflies, ants, bees and wasps.

hypermetamorphosis — an extreme form of metamorphosis undergone by some parasitic insects.

hyperparasite — (alternative term, secondary parasite) — a parasite which attacks another insect which is itself a parasite.

hypopharynx — a structure arising from the front surface of the labium in insect mouthparts and which acts as a tongue.

hypopus — a special stage in the life cycle of some mites which is adapted for clinging to and being dispersed by other organisms.

IGR — see insect growth regulator.

imago (pl. imagines) — the adult sexually mature form of an insect.

immature stage — any stage of an insect prior to the imago (adult form).

incompatible (as applied to pesticides) — not safe to mix together.

incomplete metamorphosis — (alternative terms, partial metamorphosis, gradual metamorphosis) — metamorphosis in which there is a gradual and often not very marked transformation from immature to adult stage. There is no pupal stage and immature stages are referred to as nymphs. cf. complete metamorphosis.

indigenous — occurring naturally in a country or area.

inoculation — (alternative term, introduction) — (as applied to biological control). The introduction of new species of parasites, predators or disease organisms for biological control purposes.

inhalation toxicity — the toxicity of an insecticide (to higher animals) when inhaled in the form of fine droplets or vapour.

insect growth regulator (IGR) — group name applied to chemicals which interfere with normal insect growth or development, usually by affecting growth hormone systems.

insect pathology — the study of insect diseases.
insecticide — a chemical which is toxic to insects.
instar — the form of an insect between moults.
integrated control — (alternative terms, pest management, IPM) — the use of two or more different control measures together in an integrated fashion.
integument — the outer body wall of an insect or other animal.
introduction — see inoculation.
invertebrate — any animal without a backbone (spinal column).
IPM — abbreviation of integrated pest management.
isomer — a particular molecular arrangement of a chemical compound. Different isomers of the same substance have the same atoms within the molecule but their spatial arrangement differs.
Isoptera — the order of insects comprising the termites.
JH — an abbreviation of juvenile hormone.
juvenile hormone — the hormone which retains an insect in its juvenile form; secreted by the corpora allata.
K-strategist — an ecological term applied to organisms which are adapted to ensure good survival. cf. r-strategist.
key factor (as applied to population dynamics) — a factor which is dominant in regulating the numbers of an organism.
key pest(s) — the serious and persistent pests which attack a crop.
labium — the hind or lower lip of an insect's mouthparts.
labrum — the front or upper lip of an insect's mouthparts.
lacinia — the inner arm of the "pincers" present on the maxilla of biting insect mouthparts.
larva — the active immature stage of insects which undergo complete metamorphosis. cf. nymph.
LD_{50} — dosage of chemical which is lethal to 50 per cent of the test animals treated; usually expressed in mg per kg body weight.
Lepidoptera — the order of insects comprising the butterflies and moths.
lorum — a basal sclerite present in the mouthparts of insects and well developed in bees.
low volume spraying — the application of pesticides in concentrated form using about 225 to 560 litres per hectare (20-50 gals per acre).
male confusion technique — method of disrupting mating in insects by artificial release of excess amounts of sex attractant chemical.
Mallophaga — the order of insects comprising the biting or bird lice.
Malpighian tubes — thread like organs opening into the beginning of the hind-gut of insects; excretory in function.
mandible(s) — the second pair of appendages that make up the

mouthparts of insects; in the form of biting jaws in biting/chewing insects.

mandibulate — of biting/chewing pattern.

mantle — the saddle-shaped swelling on the back of a slug.

maxilla (pl. maxillae) — one of the third pair of appendages that make up the mouthparts of insects; usually complex in structure and bearing a sensory palp in biting/chewing insects but needle-like in piercing/sucking insects.

mechanical control — any means of pest control which involves mechanically trapping or killing insects, or providing barriers to prevent insects gaining access to plants or other materials.

Mecoptera — a small order of insects comprising the scorpion flies.

mentum — the central portion of the labium in insect mouthparts.

mesothorax — the middle segment of the insect thorax.

Metabola — (alternative term, Pterygota) — the major subclass of insects which undergo metamorphosis whether complete or incomplete and which usually possess wings in the adult state. cf. Ametabola.

metamorphosis — the change in bodily form which most insects undergo as they develop. See also complete metamorphosis, incomplete metamorphosis.

metathorax — the hind segment of the insect thorax.

MH — (alternative terms, moulting hormone, ecdysone) — an abbreviation of moulting hormone.

microbial control — the use of pathogenic micro-organisms to control pest species.

micro-organism — a group term applied to living organisms that are in most stages too small to see with the naked eye; includes bacteria, viruses, fungi and protozoa.

minimum tillage — techniques of sowing crops which involve little or no disturbance of the soil.

mist blower — (alternative term, air blast sprayer) — a sprayer which uses a stream of air to carry spray droplets on to the object being sprayed.

mite — a member of the class Acari related to but distinct from the insects. Mites may be distinguished from insects by the single body region, four pairs of legs and absence of antennae.

miticide — (alternative term, acaricide) — chemical used for control of mites.

mode of action (as applied to insecticides) — the mechanism by which an insecticide kills an insect.

Mollusca — the phylum of animals to which slugs and snails and their relatives belong.

monocropping — (alternative term, monoculture) — the practice of growing the same crop (and often the same cultivar) on the same land year after year. In some situations it may involve the same crop plant over large areas.

monoculture — see monocropping.

monophagous — feeding on a single species of plant (or animal) or at least a group of closely related species. cf. polyphagous, oligophagous.

morphology — the study of the bodily form of living organisms.

moult, moulting — (alternative term, ecdysis) — the process of shedding the skin (exoskeleton) undergone periodically by insects and other arthropods.

moulting hormone — (alternative terms, MH, ecdysone) — the hormone in insects which initiates the moulting process.

mouth hooks — the paired hardened portions of the mouthparts of fly larvae which protrude from the mouth and enable them to shred their food.

multivoltine — having several generations per year. cf. univoltine.

Myriapoda — a class of arthropods which includes millipedes and centipedes.

natural control — the natural regulation of populations of living organisms that takes place without human interference.

natural enemy — any living organism which is harmful to a species by way of its predatory, parasitic or disease inducing habit.

narrow spectrum (as applied to insecticides) — having an effect on only a narrow range of insects.

necrosis — death of a defined area of plant tissue.

nectar — a sugary secretion produced by flowering plants which is highly attractive to insects.

nectary —gland from which nectar is produced; normally located within the flowers but sometimes external to them.

nematode — (alternative terms, eelworm, roundworm) — unsegmented smooth bodied worm belonging in the phylum Nematoda; mostly minute in size.

nematology — the study of nematodes.

Neuroptera — the order of insects that includes lacewings and ant lions.

nocturnal — active by night.

non-preference — a mechanism of plant resistance to pests whereby the pest chooses not to feed (or oviposit) on the resistant plant in comparison with susceptible ones.

non-target organism — any organism other than that against which control measures (especially chemicals) are applied.

nymph — the juvenile stage of an insect which undergoes incomplete metamorphosis. cf. larva.

obligate parasite — a parasite which is obliged to follow a parasitic mode of life and which cannot exist in any other way.

occasional pest — one which reaches significant levels only occasionally.

ocellus (pl. ocelli) — a simple eye of insects and other arthropods incorporating a single lens.

Odonata — the order of insects comprising the dragonflies.

oesophagus — the tubular forepart of the gut between the mouth and crop.

oligophagous — feeding on a limited range of plants (or animals), usually within a single family. cf. monophagous, polyphagous.

ommatidium (pl. ommatidia) — a unit of the insect compound eye.

omnivorous — feeding on many different kinds of both plant and animal food.

ootheca — an egg case or egg capsule of distinct form; as produced by cockroaches for example.

OP — abbreviation of organophosphate.

opisthosoma — the hind body region of arachnids (spiders, mites, ticks).

oral — via the mouth.

oral toxicity — the toxicity of an insecticide (to higher animals) when ingested.

order — a division of a class or sub-class composed of a group or groups of related families. The insect orders are the main insect types, eg, order Coleoptera, beetles; order Lepidoptera, butterflies and moths.

organochlorine (insecticide) — (alternative term, chlorinated hydrocarbon) — a group name applied to insecticides in which chlorine forms an important part of the molecule; DDT and lindane are examples.

organophosphate (insecticide) — a group name applied to insecticides which have a phosphorus atom at the core of the molecule; malathion and diazinon are examples.

Orthoptera — the order of insects that includes crickets, grasshoppers and locusts.

ostia (s. ostium) — valves by which blood enters the insect heart.

ovariole — a unit or egg tube of the insect ovary.

ovary — the female internal reproductive organ within which the eggs are produced.

oviduct — tube leading from the ovary and into which the eggs are discharged.

oviposition — the act of egg laying.

ovipositor — specialised egg laying organ possessed by most female insects.

palp — sensory feeler in the mouthparts of insects. There are usually two pairs of palps; one pair (maxillary palps) attached to the maxilla, and one pair (labial palps) attached to the labium.

papilla — small, raised, finger-like process usually of minute size.

paraglossa — the outer part of the extremity of the labium in insect mouthparts.

parasite — an organism which lives in or on another.

parasitoid — a term sometimes applied to insects which parasitise other insects (to distinguish them from parasites of vertebrate animals).

parthenogenesis — reproduction without mating.

partial metamorphosis — (alternative terms, incomplete metamorphosis, gradual metamorphosis) — metamorphosis in which there is a gradual and often not very marked transformation from immature to adult stage; there is no pupal stage and immature stages are referred to as nymphs.

pathogen — an organism capable of causing disease in a plant or animal.

penis — the male copulatory organ.

peritrophic membrane — a membrane produced within the mid-gut of insects and which encloses the food as it passes along the alimentary canal.

persistence — the degree to which the activity of an insecticide remains after it is applied.

pest management — (alternative terms, integrated control, IPM) — an approach to pest control that blends biological and chemical control methods and involves application of insecticides according to needs, rather than according to a set programme.

Phasmida — the order of insects comprising stick and leaf insects.

pheromone — a chemical substance produced by an insect or other animal which conveys a message to other members of the same species and usually results in a behavioural response.

phylum (pl. phyla) — major division of the animal kingdom corresponding to the main animal types, eg, phylum Arthropoda — arthropods, phylum Mollusca — molluscs.

physical control — pest control which involves either modification of some physical factor in the environment, (eg, temperature) or use of a physical factor (eg, light) to attract insects.

phytophagous — plant feeding.

phytotoxic — damaging to plant life.

plant cover — the degree to which plants within a sprayed area receive a deposit of pesticide.

plant host range — the species of plants on which an insect species feeds.

plant resistance (to pests) — the ability of a plant to grow and produce economically despite the presence of a pest.

Plecoptera — the order of insects comprising the stoneflies.

pollination — transfer of pollen from the male parts of flowers (anthers) to the female receptive organ (stigma).

polyembryony — the production of several (often many) embryos from one egg.

polyphagous — feeding on a diverse range of plants (or animals). cf. monophagous, oligophagous.

population dynamics — the study of changes in populations of organisms and of the reasons for such changes.

potential pest — a pest which has the potential to be serious but which is usually suppressed by natural regulating factors.

predator — an insect (or other animal) that captures and feeds on other animals (prey).

pre-oviposition period — the period between mating and the commencement of egg laying.

primary parasite (as applied to insects) — one which is parasitic on a plant feeding, predatory or scavenging species but never on another parasite.

proboscis — (alternative term, rostrum) — the extended tubular mouthparts of an insect; particularly of the piercing/sucking pattern.

prosoma — the foremost body region of arachnids (spiders, mites, ticks) to which the walking legs are attached.

prothoracic gland — a gland situated in the fore-part of the insect thorax which secretes moulting hormone.

prothorax — the front segment of the insect thorax.

protonymph — the second instar in the development of mites and ticks. cf. deutonymph.

Protura — a primitive order of small wingless insects for which there is no common name.

pseudopods — false legs on the abdomen of the larvae of butterflies, moths and sawflies.

pseudo-tracheae — fine tubes which traverse the pad-like foot of the sponging mouthparts of flies.

Psocoptera — the order of insects comprising the book-lice.

Pterygota — (alternative term, Metabola) — the major subclass of insects which undergo metamorphosis whether complete or incomplete and which usually possess wings in the adult stage. cf. Apterygota.

pupa (pl. pupae) — (alternative term, pupal stage) — the stage in the life cycle between larva and adult in insects which undergo complete metamorphosis.

puparium (pl. puparia) — specialised form of pupa in which the last larval skin is retained as an extra covering; characteristic of the higher Diptera.

pyrethroid — class of insecticides with chemical structure similar to pyrethrum.

queen — female sexually reproductive form of social insects such as bees.

r-strategist — an ecological term applied to organisms which produce large numbers of offspring but with poor survival. cf. K-strategist.

race — a naturally occurring group within a species usually visually indistinguishable but with some physiological difference from other members of the species, eg, with a different plant host range. cf. strain.

radula — the rasp-like tongue of slugs and snails.

rectum — the hindmost portion of the gut of insects (and other animals)

repellent — a chemical or physical source which induces insects to move away from it. cf. attractant.

reproductive potential — (alternative term, biotic potential) — the maximum possible rate of increase of an organism in the absence of any limiting factors.

residue — that portion of a pesticide which remains on or in produce at harvest.

resistance (of insects to insecticides) — the developed ability of a strain within an insect species to withstand an insecticide to which it was formerly susceptible.

resistance (of plants to insects) — see plant resistance.

resurgence — the recovery of pest populations (sometimes to levels higher than before treatment) following application of an insecticide.

retinula — the lower light-sensitive region of an ommatidium in the insect compound eye.

rhabdom — the dense rod-like structure lying below the crystalline cone in each ommatidium of the insect compound eye.

rostrum — (alternative term, proboscis) — the extended forepart of the head of an insect consisting of or bearing the mouthparts. The term is applied to insects such as weevils in which biting mouthparts are at the tip of a snout as well as to insects with piercing/sucking mouthparts.

saprophagous — feeding on dead or decaying organic matter or on fungi.

scavenger — an insect or other animal that feeds on dead and decaying plant and animal remains.

sclerite — a hardened plate forming part of the exoskeleton of insects.

secondary parasite — see hyperparasite.

segment(s) — repeating body units of insects and of many other animals.

selective insecticide — one which is effective against only a narrow range of insects.

self pollination — transfer of pollen from anthers to stigma within the same flower.

seminal vesicle — a storage reservoir for sperm in the reproductive system of male insects; usually paired.

sensillum — minute sensory structure on which insects' senses of smell and taste are based.

seta (pl. setae) — slender body hair or filament.

sex attractant — (related term, sex pheromone) — a volatile chemical substance produced by one sex of an insect (nearly always the female) to attract the opposite sex.

sex pheromone — (related term, sex attractant) — a volatile chemical substance produced by one sex of an insect which produces some specific reaction in the opposite sex. Most sex pheromones are powerful attractants but male pheromones may act as aphrodisiacs.

sex ratio — the ratio of males to females in a population.

Siphonaptera — (alternative term, Aphaniptera) — the order of insects comprising the fleas.

Siphunculata — (alternative term, Anoplura) — the order of insects comprising the sucking lice.

social — living together in complex colonies; eg, ants, bees.

soldier — a caste present in some social insects (especially termites and ants), whose task is to defend the colony.

solitary — occurring singly or as mating pairs; not in colonies.

soluble powder — an insecticide formulation which dissolves in water to form a solution.

species — the lowest unit of classification normally used for plants and animals.

spectrum of activity — the range of insect species against which an insecticide is active.

spermatheca — a storage organ for sperm within the reproductive system of female insects.

spermathecal gland — a gland associated with the spermatheca in the reproductive system of female insects.

spermatophore — a package enclosing sperm which is transferred to the female during copulation.

spinner — the fully mature adult form of mayflies after moulting of the dun (sub-imago) has taken place.

spinneret — the central silk spinning bristle present in the mouthparts of caterpillars (larvae of Lepidoptera).

spiracle — an external opening of the tracheal (respiratory) system of insects.

stereomicroscope — a microscope with dual eye piece and objective lens systems which provides stereoscopic vision.

sterile male technique — a method of pest control which involves the release of large numbers of sterile male insects.

stigma — a dark mark near the fore-margin of the wings of many insects; a spiracle. Also (botanical), the female receptive part of a flower.

stimulant — a chemical or physical factor which increases some specific insect activity (particularly feeding or oviposition). cf. deterrent.

stipes — a basal sclerite forming part of the maxilla in the mouthparts of insects.

stomach poison — an insecticide which has to be ingested by the insect before it exerts significant toxic action.

strain — a group within a species that is in some way physiologically different from other members of the species, eg, resistant to an insecticide. cf. race.

Strepsiptera — a small order of insects parasitic on other insects and commonly known as stylopids.

stylet — a needle-like part of the mouthparts of insects and mites that feed with a piercing/sucking action.

stylophore — the part of the head (capitulum) of mites that bears the stylets.

sub-family — a taxonomic grouping below the level of family. Insect sub-family names always end in -inae, eg, Melolonthinae.

sub-imago — (alternative term, dun) — winged but sexually immature mayfly adult which has to undergo another moult.

submentum — the basal portion of the labium in insect mouthparts.

sub-oesophageal ganglion — the ganglion (swelling) of nervous tissue below the oesophagus in insects.

sub-order — a taxonomic category below the level of order, eg, sub-order Homoptera within the order Hemiptera.

sub-species — a taxonomic category below the level of species.

super-family — a taxonomic category immediately above that of family level.

swath — the path treated by a single passage of a sprayer or other application equipment.

symbiosis — association of two dissimilar organisms to their mutual advantage.

symphilid — a member of the order Symphyla within the class Myriapoda.

Symphyla — an order of the class Myriapoda comprising the symphilids.

Symphypleona — a sub-order of springtails (Collembola) which are globular in form and without visible segmentation. cf. Arthropleona.

synapse — branched connections between one nerve cell and another.

synergist — a chemical which, when added to an insecticide, increases its effectiveness, but which is not itself toxic.

systematics — (alternative term, taxonomy) — the study of the naming and classification of living organisms.

systemic (as applied to insecticides) — having the property of being absorbed into and translocated by the sap stream of a plant or the bloodstream of an animal.

target — object onto which a pesticide is to be deposited by spraying or other means.

target cover — the degree to which the crucial portion (target) of a plant (or other sprayed object), receives a deposit of insecticide.

tarsus (pl. tarsi) — the extremity of the insect leg and forming a "foot"; composed of one to five joints.

taxonomy — (alternative term, systematics) — the study of the naming and classification of living organisms.

tegmina (s. tegmen) — the leathery fore-wings of insects such as grasshoppers or cockroaches.

teneral — newly moulted, pale and soft bodied.

teratogenic — including birth deformities.

terrestrial — living on dry land.

tertiary parasite (as applied to insects) — one which is parasitic on a secondary parasite.

testis — the male internal reproductive organ which produces sperm.

thorax — the middle of the three body regions of an insect.

Thysanoptera — the order of insects comprising the thrips.

Thysanura — the order of insects that includes three-pronged bristle tails such as silverfish.

tick — a member of the order Acari within the class Arachnida; ectoparasitic on higher animals.

tolerance (as applied to pesticides) — legally defined maximum permitted level of a pesticide on or in produce at harvest.

tolerance (as applied to plant resistance to pests) — a mechanism of plant resistance to pests whereby the plant is able to grow and produce despite pest injury.

trachea (pl. tracheae) — an air filled tube which forms part of the respiratory system of insects.

transovarial — transmission via the egg from mother to offspring.

tribe — a unit of classification below the family level into which groups of like genera are placed. Insect tribe names always end in -ini, eg, Scarabaeini.

Trichoptera — the order of insects comprising the caddis flies.

ultra low volume spraying (ULV) — the application of pesticides in extremely concentrated or undiluted form using 50 litres or less per hectare (5 gal per acre).

ULV — see ultra low volume.

univoltine — having one generation a year. cf. multivoltine.

varietal control — the use of varieties (cultivars) of cultivated plants which are resistant to pests.

variety — a loosely defined taxonomic category below the level of sub-species. Usually possesses some visibly different character.

vas deferens — the tube leading from the testis down which sperm passes.

vector — an insect (or other animal) which transmits a disease causing micro-organism from one plant or animal to another, eg, anopheline mosquitoes are vectors of malaria.

vein(s) — the strengthening ribs in the wings of insects.

ventral — the lower (front) surface of an animal. cf. dorsal.

ventral tube — a tubular protrusion from beneath the first abdominal segment of springtails.

vertebrate — higher animal possessing a backbone, eg, mammals, birds, fish.

vestigial — greatly reduced and non-functional.

viviparity — the condition of giving birth to active young rather than laying eggs.

waiting period — the prescribed period between last application of a pesticide to a crop and harvest.

wettable powder (WP) — an insecticide formulation consisting of finely ground solid particles of active ingredient plus inert carrier; forms a suspension when added to water.

wind pollination — pollination in which pollen grains are carried by the wind rather than by insects.

withholding period — the period of time following application of a pesticide, that livestock must be kept out of a pasture or fodder crop.

worker(s) — sterile individuals in the nests of social insects which forage for food and generally run the colony.

WP — see wettable powder.

yield forming organ — that part of a plant which gives rise to the harvested yield.

Zoraptera — a small order of tropical insects related to the book-lice for which there is no common name.

Index